# Retooling

*Rosalind Williams*

# Retooling
A Historian Confronts Technological
Change

Rosalind Williams

The MIT Press
Cambridge, Massachusetts
London, England

© 2002 Massachusetts Institute of Technology

Set in Sabon by UG / GGS Information Services, Inc. Printed and bound in the United States of America.

Library of Congress Cataloging-in-Publication Data

Williams, Rosalind H.
Retooling : a historian confronts technological change /
Rosalind Williams.
p. cm.
Includes bibliographical references and index.
ISBN 0-262-23223-5 (hc. : alk. paper)
1. Technology—United States—History. 2. Technological innovations—United States—History. I. Title.

T173.8.W55 2002
303.48'3'0973—dc21                                2001060365

in memory of the past: to Grandpa Lewis
with hope for the future: to Laurel, Owen, and Peter

# Contents

# Preface

In the fall of 1994, Arthur Smith, a professor in MIT's Department of Electrical Engineering and Computer Science, announced that he would be stepping down from his position as Dean for Undergraduate Education and Student Affairs. In the early spring of 1995, I got an after-dinner call at my home from Linn Hobbs, a professor in the Department of Materials Science and Engineering, who was chair of the search committee charged with reviewing the dean's office and recommending candidates to succeed Art. Would I be interested? "No, thank you," I said. I had just been asked to serve as head of the Program in Writing and Humanistic Studies, I had a book I wanted to write, and (though I did not say this to Linn) I did not want to turn over my life to such a demanding job.

Linn said he regretted this but understood, adding "Anyway, it probably wouldn't make sense to have a humanist be dean of undergraduate education at MIT" or words to that effect. Of course this got my back up. "That's the problem with MIT," I told Linn. "We still think of ourselves as an engineering school. We need to be less of an Institute and more like a university. As for leadership here, look at Jim Killian! He was one of MIT's greatest presidents, but he didn't have a doctorate and wasn't a scientist or engineer. He was a journalist, but he understood science and engineering better than many people who practice them." Or words to that effect.

Only much later did I realize how wily Linn had been in pushing my buttons. I had reacted with almost embarrassing predictability. He asked me to come and tell the search committee what I had told him, not necessarily as a candidate but to provide perspective. And so it goes: one thing leads to another, your competitive spirit gets engaged, you end up drawn into the search process, you start imagining what it would be like to do a demanding but interesting job.

Nevertheless, when President Charles Vest actually offered me the position in early June 1995, I was anxious, bordering on terrified. I was not really afraid that I would fail at it; I was afraid of being consumed by trying to do it well. As I was wrestling with deeply conflicting emotions, this thought came to me: I could get an interesting book out of this experience. As a cultural historian of technology, I would find the dean's position more than an administrative responsibility. It would also be a form of research, providing a vantage point from which to observe contemporary interactions of technology and culture. I did not exactly relax, but I felt better about what I might be getting into. I accepted the job, bought a notebook, and started recording events and impressions for this undefined book.

Seven years later, here it is, the record of a participant-observer. I have no illusions that this book has any authority as an ethnographic study; I respect my colleagues who are anthropologists too much to make any such claims. Instead, I write as a historian trying to figure out contemporary events and trends as part of larger patterns. Henry Adams said that, while history is "in essence incoherent and immoral" and thus cannot be ordered as a system, it can be written as a catalog, a record, a romance, or an evolution (*The Education of Henry*

*Adams: An Autobiography*, Houghton Mifflin, 2000 [1918]),
pp. 300–301). As Adams himself demonstrated with incompa-
rable mastery, history can also be written as autobiography as
an individual tries to see how history works through the lens
of personal experience.

A striking characteristic of the 1990s was the extent to
which both personal and historical experience became dominated by technology. I began to notice this at the outset of the
decade, when suddenly, it seemed, dinner-party conversations
with friends began veering toward computers: what games
your kids were playing, what software you were using, and
soon what you were doing on the Internet. At the outset of the
next decade, these conversations had taken on a more ominous
edge: When would technology stocks recover? Would the airline industry ever recover? Could technology be protected from
terrorism? In this book I have tried to combine such personal
experiences with a range of MIT experiences—new software
systems, new student activities, new ways of teaching, building
projects, curricular reforms, corridor conversations, committee
conversations—in order to understand how history works in
an age dominated, if not determined, by technological change.

# Acknowledgments

In a list of acknowledgments, it is usual to end by thanking one's family. For this book I need to start with them, for reasons that will soon become evident. One of my grandfathers gets considerable attention in this book, but in order to keep things focused I give much less attention than I should to others who have come before me—notably Grandma Lewis, who was such an important part of my grandfather's life and my own; my uncles, each of them so accomplished in such different ways; my mother, who majored in and taught math in a generation where women typically did not do these things; and my father, who worked as a General Electric engineer for over forty years, and whose life story is an unforgettable reminder of the value of the engineering profession as a route of upward mobility. My indebtedness to my husband and three children goes well beyond the usual tribute to their patience and understanding as I was writing this book. That was nothing compared to their patience and understanding during the five years I served as dean. In demanding jobs, the price is often paid less by the job-holder than by the people closest to her.

At the same time—life is so complicated—I must also pay tribute to the support I have received from my extended family, as it were, at MIT: faculty, students, and especially employees who worked in the dean's office. The people who work at MIT created this book, and I hope to honor them in it. One of its themes is that technological systems are always human systems. While I could not begin to list the names of the staff members who helped me so much during my years of service as dean, I hope that in these pages, at some point, they may recognize their labor and their voice.

A number of past and present MIT leaders—among them Chuck Vest, Bob Birgeneau, John Hansman, Nancy Hopkins, Lorna Gibson, and Bob Weatherall—read the manuscript and commented on its accuracy and fairness. In reviewing the manuscript, they were, in a sense, also being asked to review an institution to which they have given their hearts, minds, and energies. Nothing more confirms my respect for MIT than their generous and thoughtful response to this task.

I am also grateful to John Staudenmaier, S.J., who encouraged me to submit a much-condensed early version of what became this book to *Technology and Culture*, the journal of the Society for the History of Technology. John also read and commented on the book manuscript, as did my friend and colleague Thomas P. Hughes. Tom's reading was uniquely insightful because he had known my grandfather when Tom was an assistant professor at MIT many years ago. In fact, it was through my grandfather than I ended up working as a research assistant for Tom while I was still in college—the beginning of an intellectual and personal association with Tom and his late wife Agatha that I treasure.

The person most directly responsible for the publication of this book is Larry Cohen, editor-in-chief at The MIT Press. From the start he has understood, often better than I have, what I have been trying to accomplish. He guided me through the writing and review process, knowing when to leave me alone and when to make suggestions. In the editing process, I had the pleasure of working again with Paul Bethge. I was again reminded of the immeasurable value of a smart, sympathetic editor. The role of The MIT Press in bringing this book to completion is another reminder that among the many gifts of my life—my family, living and dead, friends at MIT and elsewhere—one of the most precious is my lifelong association with a great institution.

# Retooling

# 1

## Living in a Technological World

### Spring Garden Farm

My cousin Mary, who practices family therapy and studies
Jungian psychology, tells me that some dreams must be obeyed:
they command you. In December 1994 I had such a dream.
When I woke up I could not remember its details, but I knew it
was about my maternal grandfather, Warren Kendall Lewis,
and I knew that I had to visit Spring Garden Farm, near the
town of Laurel in southern Delaware, where he had grown up
in the 1880s and the 1890s.

Throughout his long career as a chemical engineer, Grandpa
Lewis thought of himself as a farm boy. As a professor at
the Massachusetts Institute of Technology, he often would try
to get some engineering point across by telling a funny story
about a country schoolteacher or about an apocryphal Cousin
Willie. A teacher's job, he said, was to "get the fodder down
on the ground where the calves could get at it."

In the 1960s I spent a summer with Grandpa and Grandma
Lewis at their home in Newton, Massachusetts, helping my
grandfather with research for a speech he was writing on
"Man's Use of Energy." On a lovely morning, after finishing
breakfast, we went into the sitting room to talk about the

project. We settled in chairs by an open window. We could see raindrops on the rhododendron leaves outside, beads scattered by an overnight thunderstorm. "When I was a kid on the farm," my grandfather said, "I always knew when it had rained during the night, because in the morning my father would sing so loudly as he walked to the barn to feed the animals."

Warren Kendall Lewis (1882–1975) was an only child. Both he and his parents expected him to inherit and run Spring Garden Farm, which had been in the family since the 1700s. They raised a variety of crops (fruits, vegetables, grains, fodder) at a time when mechanization was just beginning to reach southern Delaware. There was one steam-powered thresher in the Laurel area, hauled from farm to farm by horses or mules.

Warren and his parents thought he could do a better job of running the farm if he had a college education in the agricultural arts. Laurel had a three-year high school run by a principal who taught all the subjects for all three years. The highest subject in mathematics was algebra, taught in the senior year. My grandfather failed. The Lewises understood that one could not get into college from a high school like that, and so they decided to send Warren to a better school. One of Warren's cousins, Mary Witherbee, had moved from Laurel to Newton, where she taught English at what was then the Lasell Female Seminary. Orphaned at an early age, she had lived with the Lewis family. Her fiancé had died suddenly, and she doted on Warren, becoming like a second mother to him. In the fall of 1898, at the age of 16, he moved to Newton be near his cousin Mary and to attend Newton High School, which had a national reputation for excellence. During the winter of 1898 he boarded with a Newton family. On Wednesday, February 15, he wrote in his diary: "I went down town in the evening and ride. I asked a man passing for a ride and he consented; it was the first time I've ridden on or behind a horse since last Sept., and I a farmer's boy." But things were changing back on the farm too. In a letter from his father that spring, Warren learned (as he noted in his diary) that "four factories were going up in Laurel and vacinity for canning, he is going to put out 10 acres of tomatoes, he has only four cows now, etc."

In his first quarter at Newton High, my grandfather got an A in arithmetic, a B in geometry, and C's, D's and F's in everything else. By the second quarter he got five A's, two D's (one in Latin, one in drawing), and a C (in German). His plan was still to study agronomy so he could use what he learned back on the farm. On February 28 he wrote in his diary: "Yesterday I asked Miss Macumber and Miss Constantine about what they thought would be best for me to do about going to college. They will think it over more but the former advised either Amherst Agricultural or as even better Tech. Miss Constantine mentioned Cornell. I will write there. She also added I ought to have 9 hours sleep per day, a great deal of exercise and less A's. She thought I may have had 3 A's to many."

By then Warren was boarding with the family of a high school friend who planned to go to Tech, or Boston Tech (then common names for the Massachusetts Institute of Technology). He felt he had learned from his experience at Newton High that he needed to be at a competitive place. Boston Tech was supposed to be hard. He enrolled as a freshman in the fall of 1901.

He always said that he had planned to go back to the farm, but that things had just kept happening to him. It was not really so accidental, of course. He was driven by a desire for education and by a desire to excel—to get the A's, despite what Miss Constantine said, and to go to a place that was hard. Since MIT did not have agronomy, he majored first in mechanical and then in chemical engineering. Most of his tuition came from the sale of tomatoes to the canneries around Laurel for $5 a ton. Summers he returned to Delaware to work on the farm. One summer he came up with the idea of feeding cows the peels removed from tomatoes before canning. The experiment failed: the peels made cheap fodder, but the cows gave pink milk.

In February of his senior year, Warren Lewis fully expected that in June 1905 he would be getting on the train to return to Delaware. Early in the spring semester, however, he was offered a couple of lab assistantships, and then a full-fledged job in the chemical engineering lab of Professor William Walker. Walker then encouraged him to apply for an international graduate fellowship in physical chemistry. The fellowship took Warren Lewis to Breslau (then one of Prussia's leading cities), where in 1908 he was awarded a doctorate in physical chemistry. At this point he realized he was not going to return to the farm, at least not soon. The "farmer's boy" was hijacked by the twentieth century.

After working for a year at a tannery in Merrimack, New Hampshire, Warren Lewis returned to MIT as a professor. During World War I he was in charge of American research for gas defense. (After his death, when my mother and I were cleaning out the coal cellar in the Newton house, we found a collection of Allied gas masks, one from each country.) After the war, he and William Walker collaborated in developing the fledgling discipline of chemical engineering. With their colleague Walter McAdams, they wrote the textbook that defined the new discipline, organizing the curriculum around the principle of unit operations.

In 1920, MIT established a separate Department of Chemical Engineering, with Warren Lewis as its first head. He served in that position for nine years, but he much preferred teaching to administration. He was famous as a teacher—infamous, to some students, for the way he relentlessly cross-examined them in class, often showing them that they were "damned dumb-bells" for not seeing a point he considered obvious. Some students were afraid and resentful. Others discovered confidence and creativity.

Warren Lewis consulted extensively, notably for Goodyear Tire and Rubber and for Standard Oil of New Jersey. In the late 1930s, he and his colleagues developed a catalytic method of "cracking" that could be used to produce high-octane fuel. The first full-scale fluid-bed catalytic cracking unit went into operation in 1942, producing aviation fuel that enabled Allied planes to outrun and outmaneuver German ones in the Battle of Britain. Grandpa told me how tankers from the refineries in Venezuela docked in Liverpool and pumped fuel directly into the tanks of Spitfire fighters.

He hardly ever talked about his role in the Manhattan Project. When the National Defense Research Committee was organized in 1940, he became a vice chairman in the chemistry division. He served on the committee that made the crucial decisions about prioritizing the four industrial-scale processes proposed for separating U-235 for the atomic bomb. In 1942, at the age of 60, he took a train to Chicago to witness the first atomic chain reaction. (At the last minute, he gave up his place to a younger member of the group.)

During World War II he traveled constantly. My mother, on one visit to her parents' home in Newton, noticed his absence and asked her mother where "Pop" was. She later told me that her mother got a "funny look" on her face and said: "Well, I'm not supposed to tell anybody, but he's somewhere in New Mexico. If I need to reach him in an emergency, I'm supposed to call a General Groves in Washington." A few years later, my parents were listening to the evening news on the radio when Lowell Thomas announced that an atomic bomb had been dropped on Hiroshima, adding that experimental work on the bomb had been done in New Mexico. My mother exclaimed "Oh, that's where Pop was!" Grandpa felt it had been necessary to build the bomb, but it haunted him. He told my mother he had feared the bomb might blow the edge off a continent. In his very old age, he sometimes worried that he could not wash the radiation off his hands.

Grandpa Lewis retired in 1948. He still talked about going back to the farm, but over the years his goals had become more elaborate. He had some vague but strongly held ideas about using Spring Garden Farm to improve race relations in southern Delaware. When the Civil War broke out, the Lewises supported the Union in a state that was almost evenly split in its loyalties. Nevertheless, my grandfather's grandfather had been a slave owner. Kendall Lewis had developed a system so his slaves could earn their freedom over the years by working for wages; when emancipated, they owned some land and could support their families. Not all the slaves were emancipated, however; quite a few of them were listed in Kendall Lewis's will along with his other property. The fact that Spring Garden Farm had been worked by slave labor also troubled my grandfather. It was never clear to me just how he thought he would use the farm to improve the lot of the Negroes (as he called them) in the Laurel area, but this was an important part of his thoughts about returning there.

Of course, it was no longer his decision only. My grandmother, a loving and gentle soul but nobody's fool, had no intention of moving from the Boston area, where her family had lived for generations, to the far end of the Delmarva Peninsula.

Then in 1949 a four-lane highway, Route 14, was built down the middle of the peninsula. "The Dual," as it is still called there, cut right through Spring Garden Farm. My mother tells me that her father decided to sell the farm when the highway was built. I am sure there were more reasons than that. Surely Grandpa realized that not only Laurel but also his life had changed too much for him to go back.

Spring Garden Farm continued to be worked by a local family, longtime friends of the Lewises, but no one lived in the farmhouse. A small black-and-white photograph taken around 1950 shows my grandfather, my uncle, and my older brother on the porch of the empty house. Not much later, after the death of my grandfather's cousin Mary, I took a trip with him and my mother to sell her house in Seaford, just north of Laurel, where she had retired. In 1960 I went with Grandpa and my mother to southern Delaware for a family reunion held at a church in Bethel, not far west of Laurel. The churchyard, where Cousin Mary's young parents are buried, lies between cornfields to the north and a branch of the Nanticoke River to the south. In his role as a famous son, my grandfather gave a short talk and said a prayer. He had never stopped trying to reconcile his faith in Jesus and his faith in science.

Since that reunion, and before my dream, I had made only one other visit to Spring Garden Farm. It was in 1977, the year after the birth of my daughter Laurel. (My uncle once told me that, in the midst of a discussion about the richness of the English language, he had asked his father what he thought was the most beautiful word in English. Without a moment's hesitation, Grandpa had answered "Laurel.") Our sentimental plans to introduce our child to her namesake town ended unhappily. We stayed in a motel on the Dual, and Laurel cried most of the night. We could not get inside the deserted farmhouse. When we looked through the windows into the living room, we saw that the floor was now mostly dirt; the floorboards had rotted away.

In April 1995, obeying my dream, I flew from Boston to Baltimore, where I picked up a rental car to drive to southern Delaware. I did not take the Dual, preferring the back roads. The west coast of the Delmarva Peninsula is built up with getaway homes for people who live in Washington and Baltimore. As I drove eastward from Maryland into Delaware, the houses got smaller and less frequent. I had the sense of going back in time. The two-lane road cut through the quiet, flat farms, each, it seemed, with a vulture on patrol. I could see vestiges of truck gardening, which had made Delaware a major producer of fruits and vegetables for Washington, Baltimore, and Philadelphia in the nineteenth century and had brought the canneries around 1900. Every few miles, a "U Pik It" stand anticipated ripe strawberries and tomatoes. But the big money was elsewhere: all along the route, chicken houses—barracks for poultry—were set far back from the road. Signs advertised chicken feed and reminded motorists that "This is Perdue country."

Approaching Laurel, the road from the west ends abruptly at the Nanticoke River, where a four-car ferry trolls back and forth. I stood by my car during the five-minute crossing, watching the river birds and grasses, amazed and delighted by this pastoral interlude. On the Laurel side of the river, an African-American family was gathered, in Sunday dress and laden with pies, headed for a church picnic on the western bank of the Nanticoke. We nodded and smiled as we exchanged riverbanks.

I drove eastward from the ferry landing on a road that runs parallel to the "river walk" of the 1700s. Sailors docking at

Bethel would walk upstream along the Nanticoke to Laurel to find a place to sleep. Sometimes they would bunk in the attic of the Spring Garden farmhouse. I slept there myself that night, not in the attic but in a four-poster bed in a second-floor bedroom, listening to the creek outside as it dribbled down to the mill pond and eventually to the Nanticoke. The farmhouse was now a bed-and-breakfast inn run by an energetic distant relation of the Lewises who had become a leader of the Laurel Historical Society. She had enrolled the house in the Registry of Historical Places to protect it from alteration or destruction, and had restored it with taste and affection.

I woke to see pink cherry blossoms brushing the bedroom window. I dressed in shorts and went downstairs to take a run, leaving by the kitchen door at the back of the house, passing by the long-sealed well, then going down the driveway next to the barn. The animals were long gone. The barn was now filled with carved wooden ducks, quilts, wrought iron tools, old signs, baskets, and other so-called antiques for sale. I headed into town. Newcomers had begun to buy and fix up the old houses, painting them in trendy pastels. I ran by the small brick building that had been the railroad station where in 1898 my grandfather had climbed on the train that took him to Boston. It now houses the Laurel Historical Society. The African-Americans of Laurel live quite literally on the other side of the tracks.

I ran as far as Laurel High School, on the outskirts of town, a splay of one-story brick rectangles plus a large two-story gym. I went into the gym and checked out the hallway exhibits of trophies and photographs of the sports teams (the latter showing black players next to white). I decided to come back later to buy a school sweater with the Laurel bulldog on it for my daughter, who was then in college. Back outside I resumed my run. I zigzagged through the streets of the town as I headed back to the farm.

In my grandfather's day, Laurel, with its shops and train station, was a busier place than the farm on its outskirts. Now the situation was reversed: activity picked up as I left the quiet town and approached the highway. Across the street from the farmhouse, the subdivided land sprouted ranch houses instead of corn and tomatoes. Just beyond, on the Dual, cars moved up and down along the mini-malls, convenience stores, furniture outlets, and supermarkets. Back of the farmhouse, just beyond the well outside the kitchen, one small field was still under cultivation. Across the field I could see the back of a K-Mart on the Dual. The woman who ran the bed-and-breakfast told me that the field had been sold to K-Mart by the farmer who had worked Spring Garden for many years after my great-grandfather had died. Soon, she said, it would be a parking lot. The field is doomed, the sound of traffic is incessant, and Spring Garden Farm is an island in the techno-commercial sea that laps relentlessly at its shore.[1]

**What This Book Is About**

For me, ending up at MIT was both a gift and a surprise. I had grown up assuming that the institute was a preserve for male engineers like my grandfather. Even when I was in college, doing summer research in the MIT libraries for my grandfather's talk, it never occurred to me that I might work there some day. The title of the talk ("Man's Use of Energy") was one small indicator of the much larger set of social norms and expectations.

Grandpa Lewis died in 1975—the year before Laurel was born, three years before I got my doctorate, five years before I first came to MIT as a research fellow and lecturer, ten years before I became an assistant professor. Twenty years after his death, I was asked to serve a MIT's Dean of Students and Undergraduate Education.[2] This too felt like a command that had to be obeyed. As a woman and a humanist, I am a member of two minority groups at MIT. Humanists are outnumbered: there are 86 senior faculty members in the School of Humanities, Arts, and Social Sciences, to which I belong, compared to 232 senior faculty members in the School of Engineering. Women are outnumbered: we make up 16 percent of the faculty overall (though just under a third of the faculty of the School of Humanities, Arts, and Social Sciences).[3] I told myself that if MIT had the courage to ask a woman humanist to serve as dean of its undergraduate program, I had to summon the courage to obey. More selfishly, I realized that I would enjoy a wonderful vantage point from which to observe contemporary history of technology—a realization that has been fulfilled many times over.

Of course I wish Grandpa Lewis could have seen this outcome. During some of my difficult days as dean, I would change my usual running route near my Newton home to go by the Lewis house on Lombard Street, only a mile away from where I now live, to try to hear his words of encouragement. On the better days, I simply wished I could talk with him about the changes at MIT since he was there. Those would be wonderful conversations. He studied science and engineering, I studied history and literature, and we converged on a common interest in the human dimensions of technology. I wish we could continue those conversations into the digital age, though I am sure that Grandpa would focus not on technological changes but on social ones. Several times he told me that if he could be reincarnated he would like to come back as a social scientist: "That's where the really interesting questions are."

During my five years as dean, I came to see my grandfather's story as part of a larger story in the history of technology and society. A combination of utilitarian calculation, personal ambition, and utopian dreaming took Warren Lewis from Laurel to MIT. He wanted to be a better farmer, to help the Negroes, to improve his life and that of southern Delaware. He believed in progress and invested his faith in science and engineering. That faith helped create a new world, one that offers vastly less hard labor and more opportunity than late-nineteenth-century farm life. But it is not the world my grandfather loved. The story has poignancy, even tragedy, because he was an engineer who, with the best of intentions, contributed to the erosion of the way of life he cherished. Chemical engineering helped fuel the age of the automobile; the Dual sliced through the farm and brought a parking lot to the back door of the kitchen. When DuPont put a factory in Seaford, the employees built ranch houses in Laurel and businesses sprang up on the Dual. Chemical engineers brought the concept of unit processes to the biochemical industry; now, that industry manufactures the feed and the antibiotics that have turned the farmlands of southern Delaware into poultry factories.

MIT produces technological innovations as predictably as Spring Garden Farm produced tomatoes. Now the techno-commercial sea is lapping at this island too. In the same way that biochemistry, DuPont, and modern highways transformed southern Delaware, the information age is transforming MIT. Innovations sent forth from the postwar world of engineering are washing back to the Institute. At one meeting on the subject of educational technologies, a faculty member sighed: "We need to catch up with the impact we've had on the world." People at MIT are discovering that the world has just as much impact on us, and that "we" are no longer who we were. MIT is famous as a place from which technological change comes. It is less famous as a place that confronts change.

This book describes the less famous MIT, where, in the words of another colleague, "We are suffering from innovation." The point of the book, however, is not to describe MIT, but to use MIT as a local example of a larger society devoted to technological innovation. This period in history is defined by the pace and extent of technological change and by our socially organized obsession with it. The implications extend far beyond the obvious economic ones. Technology-driven change also defines, in a way unprecedented in history, human desires, anxieties, memories, imagination, experiences of time, and experiences of space.

In describing this integrated, felt experience of technological change (which is wholly different from market logic), abstract analysis is of little use. Historical experience is unsystematic, situated, and particular, even if it emerges from and feeds back into larger patterns of change. The MIT experiences I will recount both emerge from and feed back into more pervasive trends and themes in information-based societies.

MIT is a good place from which to view technological change, not because the place is typical, but because innovation and its consequences are so concentrated there. I cannot imagine a more pro-technology place than MIT. At its core is a deep belief in the value of technology as the basis for human improvement, and in technological analysis as the basis for problem solving. Belief and self-interest go together. The place is packed with people who benefit from the digitized global economy as researchers, as educators, as consumers, and as entrepreneurs.

Yet the sources of technological creativity and of adaptability to change depend on resources that will never be provided by global information systems: time and space for daydreaming, unplanned encounters, wild ideas, institutional legends, inherited traditions, historical rhythms. In making these resources scarce, an institution devoted to making the world a better place may be cutting itself off from deeper sources of both technology and change. In worrying about the scarcity of these resources, people at MIT are beginning to understand, if only semi-consciously, that obsession with the technological future may be undermining the very world they value so much. They are painfully ambivalent, both excited by innovation and suffering from it.

### History as Technological Change

One dominant assumption about the information age is that it represents a "new world" of technological change that relentlessly pushes into retreat an "old world" of culturally driven resistance, much as the soldiers of the Roman Empire pushed the Celts back into the western hills of Britain. Today's conflict, we are told, is between the new economy and the old one, between digital and analog, between systems and life-world, between globalization and identity. The human race is split between dynamic "change agents" and the stagnant non-retooled, between neo-nomadic "symbolic analysts" who work with bits and a "disposable labor force that can be automated and/or hired/ fired/offshored. . . ."[4]

Such binary thinking is extended to history itself. The information revolution is leading us into a new phase of history, in which technological change will be the major and constant determinant of human life. History becomes the record of a grand struggle between the irresistible force of technology-driven "change" or "innovation" on the one hand and misguided if understandable culturally driven resistance to change on the other, with "change" and "innovation" inevitably winning.

This is supposedly an age without a sense of history, but the view of history as a struggle between technological change and technological resistance is powerful and pervasive today. Implicit rather than explicit, it is less a theory (which would imply some conscious awareness of the issues and some effort at logical consistency) than an ideology (ideologies are not especially logical or consistent).[5] It is an ideology of history in which the very word 'history' has been displaced by 'technology'. Instead of being a figure in the ground of history, technology has become the ground—not an element of historical change, but the thing itself. We have come to assume that where technology is going is where history is going, as if they are now one and the same.

The ideology of technological change has become the dominant, if implicit, theory of history in our time, even for those who never pick up a history book. In Karl Mannheim's famous distinction, ideology is a set of "wish-images" intended to evolve the existing order into a better one, while utopia represents a wish to break the bonds of the existing order, to overturn it for a new reality. Mannheim adds: "The utopia of the ascendant bourgeoisie was the idea of 'freedom.'"[6] Innovation has come to define a utopia of freedom, an ideal place running by market and technological laws, where creative minds and free-flowing capital unite to make a new world abundant in possibilities and energy. This is a much more appealing world than the one we actually live in, where innovations become hardened into bureaucracies, where actions have consequences, and where waves of change wash back on shore. The utopian "no place" of market-driven innovation presents an agreeable escape from the complexities and consequences of a crowded, risky world. Mannheim emphasizes that ideology and utopia are ideal types; in practice, it is often difficult to distinguish them within the social order or even within an individual's mind. In the view of history as technological innovation, however, we can often discern a set of "wish-images" that is utopian without being especially progressive.

As an MIT administrator, I rarely heard anyone talk explicitly about history, but I often heard people express the implicit view of history as technological change. They did this through repeated use of a cluster of deceptively simple words: 'technology', 'change', 'culture', 'space', 'time', 'community'. These words are highly interdependent, deriving their meaning from the others and forming a closed world of cross-references. They are also highly reflexive, acquiring meaning from the very changes they are supposed to analyze.[7] They also drain meaning away from other, formerly more significant words, such as 'engineering' and 'progress'.

The sun of this linguistic system is 'technology'. In 1861, when MIT was given its name, the word was novel, even daring, and grandly comprehensive. Proclaimed in chiseled stone below the Great Dome, 'technology' still defines, in a lofty way, the Institute's mission. But in everyday meetings below and around the dome 'technology' has little grandeur. It has come to mean *information technology*—a recent, drastic, telling reduction. In committee meetings, when we sketched business processes or organizational models on a flip chart, someone would write 'technology' with a circle around it. We would talk about designing new learning spaces to accommodate "the technology." We would write and review reports saying things like "The primary motivation to revise the policy [is] the recognition that technology has transformed the means by which we collect, manage, and provide access to student information."

At no time, in any of these contexts, did anyone suggest that 'technology' might include more than information technology, nor did anyone challenge its status as an independent entity. 'Technology' once was a grand term, though also a dangerous one, because it was so often used as an abstract, independent historical agent.[8] It is still dangerous in this way, but it is no longer grand. It means information technology and it means change.

While 'technology' expands its rhetorical reach, that of 'engineering' shrinks. Never a glamorous term (though it was a solid one), it is now used more rarely, and mostly in a connection with a specific project or department. Even more conspicuous in its absence is the word 'progress'. Not long ago, discussions of technology were dominated by "progress talk."[9] At MIT today, people talk a lot about scientific progress, referring to the kind of discovery that would win a Nobel Prize, but little about technological progress, except with irony.

Instead, 'technology' is linked with 'change'. Everyone talks about change as relentless and inevitable but also, almost always, as pointless, in the double sense of lacking a purpose and lacking an end. Progress has a story line; change does not. After one meeting about new software systems, a participant told me that a consultant had advised her to use the concept of a "change journey" to help people overcome their resistance to change. When I inquired what was the destination of such a journey, I was told that it might be thought of as arriving at a point of comfort with change. 'Change', like 'technology', has been sharply reduced in meaning. Of all the possible types of change—political, intellectual, cultural, social—technological change is all that really matters.[10] The change journey takes us through familiar landscapes of science, engineering, politics, the market, work, family. These too may change, everyone acknowledges, but the assumption is that they will do so only as a consequence of technological change.

Technological innovation resembles but hollows out the idea of progress.[11] Progress is measured in reference to human-defined goals: change and innovation are measured by market success. Market-driven innovation may tend to move in certain directions (Moore's Law, globalism, systems integration), but those directions are defined less by human purposes than by the rules of the market and the inner workings of technology itself. Innovation does not necessarily translate into human progress, any more than biological evolution means progress of the species (as Darwin well understood). Today's ideology of history as technological innovation turns the market, the motor of capitalist acquisition, into the motor of history itself, with technology as its agent.

It is easy to unmask the rhetoric of innovation and change, and to show that it is a cynical management tool justifying any sort of human disruption (moving, working harder or longer, being laid off) as the inevitable downside of the larger good. But then why is this view of history so widely, if semi-consciously, accepted?[12] Management babble is heavily discounted by most of us. In the midst of its ideological torrent, we try to keep living reasonably human lives, trying to figure out who we are, how to connect with others, how to accomplish something meaningful, how to enjoy the time we have. Still, the rhetoric of change management has some ring of truth, because it expresses, however inadequately, everyday experience in a world dominated by technologies we have created.

The rhetoric of change management has the effect, however, of trivializing the whole concept of change. Significant change is much more than a new software system or product. It is historical, in that it involves human relationships, expectations, and meanings. There is no single digital divide that we will step over collectively into the information age, leaving behind the creaky old world of everyday life for a sleek new world of frictionless systems. There are many fault lines in history today, and many of them run right through us. The change agent and the change resister are often the same person.[13] We are living through a transformation of the human habitat that is much more comprehensive than what is usually thought of as "technological change."

**The New Human Habitat**

Predictions about the information revolution have become the "airport reading" of the world—a substitute for beach reading in a harried age, just as predictable but a lot less engaging, celebrating not the steamy pleasures of physical reality but the disembodied pleasures of virtual reality. Computing will be ubiquitous, our rooms will read our thoughts, the air will hum with messages, the Internet will become an omninet, we will be lifted out of the mud of localism to digital globalism— all on the other side of the great historical divide. Many historians of technology dismiss most of these predictions as unfounded digital evangelism. Historians are trained to look for continuity, to be wary of revolutionary claims, to spot inconsistencies. On the latter point, Alex Roland, a recent past president of the Society for the History of Technology (SHOT), notes that people who talk about technological revolution are often talking about different things: "Fifteen years ago we were confidently told that we were in a computer revolution. Now it is an information revolution. A more proper label might be a communications revolution. Or it may turn out in hindsight to have been a solid-state-physics revolution, or a digital revolution, or a micro-electronics revolution, or a control revolution. Some people think that the coming revolution in genetic engineering will outweigh all of these."[14]

So what kind of technological change is it that we are living through, and how does it compare with other major transformations in history? Is this another industrial revolution (the productive one of steam engines, cotton mills, and railroads), or is it more like the second industrial revolution (the consumer one of electricity, automobiles, cinema, and chemical industries)? Is the telegraph "the Victorian Internet,"[15] or is the Internet more like electricity? Or should we compare information to energy?

No analogy seems quite right. The industrial revolution of the late eighteenth and the early nineteenth century was defined by an unprecedented leap in human productivity brought about by the exploitation of the vast energy reserves stored in fossil fuels.[16] If you compare the efficiency of the steam-powered thresher in Laurel with the labor of hand threshing, and if you multiply that kind of efficiency millions of times in millions of settings, you begin to grasp the essence of the industrial revolution.

The second industrial revolution (which occurred in the late nineteenth century) also increased productivity, but its distinctive contributions were the distribution of cheap and flexible energy in the form of electricity, the construction of grid upon grid of communication and transportation systems (subways, automobiles, telephone, radio), and what contemporaries called "the democratization of luxury" through the proliferation of cheaper and more varied consumer goods. Collectively these technological innovations made everyday life far more comfortable and interesting than ever before. In the words of another past president of SHOT, Robert Post: "The late nineteenth century was the greatest period of technological change in terms of things that affected huge numbers of people's lives in basic ways—much more so than today."[17]

In all these discussions, historians are trying to develop a sort of historical Richter scale: What are the minor shocks, and what are the major ones? All of us are trying to get a sense of the scale of change in the midst of constant tremors. At one MIT meeting about educational technologies, a participant from the software industry commented: "Education is the building and technology is the earthquake." On the historical Richter scale, however, many historians of technology would say that the so-called information revolution is not so big. They disagree with the hypothesis of fundamental change preached by digital evangelists, in part because historians are used to measuring change by productivity rates and living standards. These measures miss some distinctive features of information technology. At the heart of information technology is the manipulation of symbols, not of matter. It does not simply "impact" culture and society; the technology itself is inherently cultural and social from the start. Directly as well as indirectly, it affects human experiences of space, time, communication, and consciousness.

From this perspective, the most appropriate historical analogy is another communications revolution: the invention of the mass media, or the printing press, or perspective, or numerals and the alphabet, or, behind them all, the invention of language—by far the most complex and sophisticated human invention, and a prerequisite for storing and passing on all other forms of technological knowledge and practice.[18]

But this analogy too is inadequate, insofar as information technology has transformed production and consumption as well as consciousness and communications.

If comparisons are hard to find, it is not because technological change is less significant today than in the past but because we are breaking it up into episodes instead of seeing it whole. The digital divide is a major episode, not the whole story. The technological revolution we are living through includes everything Roland names: computers, information, communications, solid-state physics, digital electronics, control, genetics. It also includes the two industrial revolutions, which now may be seen as partial and preparatory. The essential feature of the larger technological revolution is the creation of a new habitat for human existence.[19] On the historical Richter scale, it is The Big One. Elting Morison, a historian who joined the MIT faculty in 1946, described it as follows:

We are well on the way, in our timeless effort to bring the natural environment under control, to replacing it by an artificial environment of our own contriving. This special environment has a structure, a set of tempos, and a series of dynamic reactions that are not always nicely scaled to human responses. The interesting question seems to be whether man, having succeeded after all these years in bringing so much of the natural environment under his control, can now manage the imposing system he has created for the specific purpose of enabling him to manage his natural environment.[20]

Morison's formulation shows how difficult it is to find adequate and precise (not to mention gender-neutral) language to describe the new human habitat. Calling our environment "artificial" or "special," as Morison does, suggests something unnatural or even abnormal about it. There is nothing more natural, however, than the human creativity and ingenuity that have shaped this new environment. For human beings, the production of technology is as natural as is the production of

enzymes for certain bacteria.[21] Nor is an "artificial" environment "replacing" the natural one. Nature has ended not in the
sense of going away but only in the sense of being so mixed
and mingled with human processes that it can no longer be
identified as a separate entity.[22] In the words of the social the-
orist Ulrich Beck: "Not a hair or a crumb of it is still 'natural',
if 'natural' means nature being left to itself."[23]

Human beings have always tried to control nature so as to
make life safer, more predictable, more abundant, and more
fulfilling. But since the beginning of recorded history, and
back into unrecorded history, non-human nature has been the
ground of human life. This relationship between technology
and nature, between figure and ground, which had been
reversing slowly over centuries, reversed decisively in the past
century. The built world has become the ground of human
existence, now framing and embedding non-human nature.
We have gone from using technology to control and exploit
our habitat to using it to detach ourselves from our habitat.[24]

We continue to struggle for words with which to express this
new phase of history. As early as the first industrial revolution,
Georg W. F. Hegel contrasted "first" or given nature with
"second nature" created by humankind. In similar language,
the historian of technology Thomas Parke Hughes has referred
to the technological world as the "second creation." In the
1960s, Lewis Mumford contrasted "the organic habitat" with
"this new megatechnics...a uniform, all-enveloping structure,
designed for automatic operation."[25] I find myself using the
verbal shorthand of referring to a technological world, or,
more accurately, a *hybrid* world—a world in which technology
and nature are inextricably mixed.

There is a convincing historical precedent for the replacement
of one human habitat by another: the Neolithic revolution,

usually dated around 10,000 BCE and usually defined by the integrated invention of agriculture, of cities, and indeed of history itself, since the concomitant invention of writing permitted, for the first time, a durable collective memory. Left behind, except in specific settings and in deepest memory, was the pastoral life of hunting and gathering, which had been the human, or humanoid, way of life for hundreds of thousands of years.[26] All the transformative events of the Neolithic revolution—writing, agriculture, settlements—were based in material change, but they also redefined what it meant to be human.[27] As the historian of religion Mircea Eliade notes, when most of humankind started raising crops "an ancient world, the world of nomadic hunters, with its religions, its myths, its moral conceptions, was ebbing away." Eliade continues: "Thousands and thousands of years were to elapse before the final lamentations of the old world died away, forever doomed by the advent of agriculture. One must also suppose that the profound spiritual crisis aroused by man's decision *to call a halt and bind himself to the soil* must have taken many hundreds of years to become completely integrated."[28]

The technological revolution of our times is the "decision"—collective, unconscious, incremental—to unbind ourselves from the soil. Eliade reminds us how far-reaching are the implications of binding ourselves instead to a self-constructed world. For millennia, the fact of settlement—humans living with other humans in a place over time—has shaped our ideas and practices of work, family, time and space, and society. This fact has also shaped our souls: solace, redemption, transcendence, and meaning have been defined by the twin powers of human community and non-human nature. It may be centuries before the lamentations for a lost world fade away and the human imagination can come to terms with its new habitat.

A technological world is partly natural and intensely human. The new fact of history on a material level is that nature is now so thoroughly intermixed with technology. The new fact of history on a social level is that we keep running into ourselves, as it were, as we build our values and our social order into the world. Sociologists describe this as the transition from modern society to something beyond: post-modernism, or (as Anthony Giddens prefers) post-modernity or radical modernity, or (as Ulrich Beck prefers) a risk society. Such thinkers stress the "reflexivity" of our society, meaning that we live in a world of echoes, a "boomerang" world (Beck's term) where everything that goes out comes back, where everything gets sampled and remixed, where everything has consequences, where social relationships co-evolve with material ones, where technology changes the very institutions producing it.[29]

The reflexivity of the world can be read positively (Giddens: "The reflexivity of modern social life consists in the fact that social practices are constantly examined and reformed in the light of incoming information about those very practices, thus constitutively altering their character"[30]) or less positively (Beck: "The gain in power from techno-economic 'progress' is being increasingly overshadowed by the production of risks"[31]). In any case, the process keeps getting more intensely reflexive. Once the marketplace takes the prime role in directing technological change, as it has since the 1970s, the process feeds back into itself.[32] A leading product of information technology is more information technology, all molded by the imperatives of a capitalist economy.

In terms of technological development, this is a virtuous circle—but it is still a circle, which accounts for the cultural paradox that the digital age engenders both a sense of liberating possibilities and a sense of oppression. Because of its socialized nature, information technology enhances social abilities, arguably our most human trait. But as information technology keeps reinforcing its dominance in terms defined by the market, other forms of sociability get selected out. A business-oriented society becomes more and more so as the system responds to its own feedback. In a world we have largely constructed, we keep encountering ourselves—but the projected "selves" we keep running into represent only a part of capacities, and arguably not the finest part.[33]

In the words of the historian Elting Morison, the historical challenge of our time is to build "a technological firmament that will really fit us" and "to organize a technological world we can live in."[34] History has not ended any more than nature is ended. Inevitably history will reveal itself to be much more powerful and unpredictable than the latest software upgrade. Inevitably nature will remind us of the fragility and impermanence of our control of the organic habitat. But history and nature are weakening as well-defined external frameworks that give meaning to human life. For most of human experience they were much more vast than the scale of individual existence, and their vastness provided solace in the case of nature and hope in the case of history. When nature becomes so intermingled with humanity, and when historical change becomes defined as an extension of technological change, neither nature nor history provides a self-evident framework for human meaning and identity. The only larger framework left, it seems, is humanity itself.

### Change and Community at MIT

"We find ourselves on the cutting edge of changes that we do not fully understand," an MIT vice president once said, speaking of the Institute. This book is an effort to close the gap between change and understanding, and not just at MIT.

During my years as dean at MIT, the economy boomed and the Internet arrived. For MIT, however, they were years of pain as well as gain. The Reengineering Project, begun in 1994 and officially ended in 1999, brought some necessary functional improvements but also brought considerable anxiety and dissension. The alcohol-related death in 1997 of a freshman living in a fraternity led to anguished questioning of the quality of student life at MIT and to a controversial and sometimes bitterly resisted decision to require all freshmen to live in MIT-operated residences. The strains on the MIT community caused by these and other episodes were not far in the background as MIT undertook a high-level reexamination of its educational mission through the work of the Task Force on Student Life and Learning, convened by President Charles M. Vest in 1996.

At fin de siècle MIT, the air was full of "change talk" and "community talk." The two discourses often contradicted each other. Change seemed inevitable, but it also seemed to undermine community. Community seemed invaluable because it seemed necessary to create and support change. The story of those years—and I suspect of years to come, and at many other places too—is the effort to reconcile change and community, future and past, in an age of technological innovation.

If reconciliation is possible, it will happen only when as much attention is given to social creativity as to technological creativity. Social creativity is necessary for a better future, but it always emerges from circumstances of the past. In the case of MIT, the past circumstances that matter most are related to MIT's institutional identity as an engineering school. Since its founding in 1861, engineering has been MIT's "legitimizing identity." Engineering has generated MIT's organizational structure, which confirms the personal and professional identity of the people there; this sense of identity, in turn, explains and strengthens the organizational structure.[35]

But now engineering itself is undergoing an identity crisis. The displacement of 'engineering' by 'technology' is just one sign of this identity crisis. The role of engineers in society is changing dramatically. Their work, which once took place place within powerful corporate and government institutions, has become more individualized as those institutions have weakened. The definition of engineering knowledge, especially in relation to scientific and managerial knowledge, is uncertain. Even more fundamentally, the mission of engineering has to change when its dominant problems no longer involve the conquest of nature but the creation and the management of a self-made habitat.

Engineers are famously hard-working and enjoy making things—in the parlance of MIT, they like to "tool." But the role of engineering in a reflexive and hybrid world is much more complex than making things. By "tooling away" with such energy, engineers restructure their own habitat. To adapt to their new surroundings, they have to retool, starting with the understanding of engineering as a profession. Today, technological change is something that happens to engineers as much as to anyone else.

were, respectively, 0.5, 1.7, and 1.4. At the end of the century, they were 7.6, 2.7, and 3.5.[3]

There is no "end to engineering" in the sense that it is disappearing. If anything, engineering-like activities are expanding. What is disappearing is engineering as a coherent and independent profession that is defined by well-understood relationships with industrial and other social organizations, with the material world, and with guiding principles such as functionality. Engineering is "ending" only in the sense that nature is ending: as a distinct and separate realm. The two processes of disintegration are linked. Engineering emerged in a world in which its mission was the control of non-human nature and in which that mission was defined by strong institutional authorities. Now it exists in a hybrid world in which there is no longer a clear boundary between autonomous, non-human nature and human-generated processes. Institutional authorities are also losing their boundaries and their autonomy.

To see the expansive disintegration of engineering, one need only look at the list of engineering departments at MIT. Their traditional boundaries are crumbling as departments stretch and reshape themselves to confront a host of problems, varying wildly in scope and character. The departmental structure developed during the nineteenth century and the early twentieth century. This development is particularly evident at MIT because MIT's departments ("courses" in MIT parlance) are numbered in sequence depending on when they were formed, the oldest departments having the lowest numbers. Course I is Civil Engineering, Course II is Mechanical Engineering, and so forth, down to Course XXII (Nuclear Engineering). During most of the twentieth century, engineering at MIT was organized in these departmental "nuclear families." Now the School of Engineering has become a collection of extended families, with stepchildren, in-laws, and hyphenated last names.

Civil engineering, once synonymous with construction, has for decades been subdividing into other activities, all eminently useful to society but sharing little in the way of a common identity. As long ago as the early 1960s, Dean of Engineering Gordon Brown reportedly asked in exasperation: "What is Civil Engineering, a holding company?"[4] Today, one side of the department deals with building technology (e.g., how to make concrete) while another side studies the management of complex construction management systems (such as how to get concrete to a site in a timely way). Indeed, around 1990 some members of the Civil Engineering faculty proposed that the department redefine itself as the primary "engineering systems" department at MIT, since so many of them worked in that area. The department did not formally assume such an identity, however. Instead, in 1992 it added 'Environmental' to its title, and since then faculty have continued to carry out long-standing research interests in hydrology and water transport. Some faculty who work on atmospheric circulation and hydroclimatology have closer intellectual ties to the Department of Earth, Atmospheric, and Planetary Sciences (which is in the School of Science) than to other members of Civil and Environmental Engineering.[5]

Similarly, the Department of Mechanical Engineering is now a conglomerate of markedly different types of research and teaching. Faculty members there use technologies from a wide array of disciplines (materials, network algorithms, electronics, magnetics, optics, computer science) to create new interdisciplinary mixes in which "mechanics" does not necessarily dominate. The department's activities range from fundamental science (for example, cryogenics and nanoscale phenomena) to teamwork-based design. "The attitude seems to be that anything that is interesting, we'll call mechanical engineering," one senior administrator said wryly. This is by no means a bad way to organize the department, but it does mean that the identity of mechanical engineering has been blurred.

Not long ago, the Department of Chemical Engineering was synonymous with chemical process industries, especially the petrochemical industry. In the 1970s, it began to shift its focus to biotechnology. MIT's largest chemical engineering lab designs body tissues (including artificial skin) and drug delivery systems. One MIT scientist remarked that the department now seems indistinguishable from a medical school or a biology department.

In some departments, the mix of science and engineering has been so well stirred that the name of the department has been changed as a result. In 1967, Mining Engineering and Metallurgy merged to become Metallurgy and Materials Science; in 1974, that department became Materials Science and Engineering. In 1975, Electrical Engineering became Electrical Engineering and Computer Science (EECS).

Other engineering departments defined by their relationship to a particular industry had to restructure their identity depending on the fortunes of the industry. With the decline of nuclear power generation, many faculty members in Nuclear Engineering shifted their interests toward medical research in what they now call "radiation science and technology." Aeronautical Engineering, established as a full-fledged department in 1939, became Aeronautics and Astronautics in 1959. Since then, as aircraft construction has declined in economic importance, the department has added a large range of non-construction-related research topics, including logistics, human engineering, and environmental impacts. For example, the Humans and

Automation Group in Aero-Astro (as that department is commonly called) does "cognitive engineering" at the intersection of human factors and systems engineering in what it calls "high information environments."

This is not the end to mutations and hybridizations in the School of Engineering. In 1998 that school added two cross-cutting divisions, the Biological Engineering Division and the Engineering Systems Division (ESD). In the new matrix organization, the divisions cut across traditional departmental boundaries so that a faculty member may have a "two-key" appointment (one in the division, one in the department). In both divisions, the intention is to recognize common approaches to engineering that cut across traditional disciplinary lines. In the Biological Engineering Division, the line between engineering and science is blurred beyond recognition. In the ESD, the line between engineering and management is increasingly hard to discern. The ESD has close ties with MIT's Sloan School of Management, where the term "financial engineering" is often heard and where Merrill Lynch and MIT have launched a multi-million-dollar collaboration in Financial Technology.[6]

Any "engineering school," and especially one as unusual as MIT, gives an incomplete and distorted representation of engineering as a whole. There are many other varieties of engineering in other educational institutions, in business, and in government. Engineering also looks very different depending on where you are in the world. As manufacturing and production have tended to move overseas, engineering in the United States has undergone a general shift, along with the rest of the economy, from an industrial to a service orientation—with the important caveat that the whole distinction between goods and services is breaking down. MIT is neither inclusive nor typical; however, because it is determined to be a world leader in engineering, it provides unusual predictive power.

In engineering, the cutting edge is a blurred edge. The more you look at its outer limits, the less you see familiar boundaries and familiar technical problems. If you look into the future, engineering seems to be evolving into new, sometimes exotic, and only loosely related life forms. If you look into the past, the identity of engineering looks considerably more coherent, though not without internal tensions. It is worth reviewing this older identity and its tensions in order to understand why the identity of engineering has been destabilized, simultaneously expanding and disintegrating, and what this means for the future of engineering.

## The Ideology of Engineering

When I was a little girl, on many holidays my family made a six-hour drive from Schenectady, New York, where my father worked for General Electric, to Newton. We would usually stay with Grandpa and Grandma Lewis, who lived in a large brown-shingled house built in the 1890s, not long before my grandfather came to Newton to attend high school. After Thanksgiving dinner, or Christmas dinner, or Easter dinner, we would all move from the capacious dining room to the cozier sitting room. The conversation there was dominated by my father, my uncles, and my grandfather, and their favorite topic was the unappreciated contribution of engineers to modern civilization.

Many times, while sitting in a meeting of an MIT committee or council, I have been transported back to my childhood as I listened to technically minded males resonating off of one another's frustrations and obsessions. They often got so wound up that they forgot I was even there. Other women faculty members at MIT have told me about similar experiences of personal time travel. As one of them said to me, the family dinner table was where you learned to deal with these guys, because you were stuck there.

As a girl, I found the conversations in my grandparents' sitting room quite interesting. They were a door into the real world of adult organizations and responsibilities at a time when there was no other such door in my life. I was alternatively amused and exasperated by the tone of these conversations. My grandfather, the alpha male, was particularly intent on defending the social contribution of engineering. Most people, he said, misunderstood what engineering had done for civilization. These "poor fools" thought that the fruits of engineering were gadgets like tailfinned cars, jet planes, ranch houses, or space ships. No, he proclaimed; the fruit of engineering, beginning with the industrial revolution, was the liberation of humankind from toil and deprivation. Only with the enormous increase in human productivity that engineering had brought could ordinary people enjoy the leisure that made it possible for them to be full participants in civilization. Being an engineer, Grandpa Lewis used numbers to make his point. When he entered Newton High School, 5 percent of the children of high school age in the United States went to high school; by the 1960s, it was 85 percent. When he went to work at the tannery in New Hampshire, people there worked 78 hours a week; 50 years later, a 40-hour week was the norm.[7]

But my grandfather, my father, and my uncles were often frustrated that the larger world failed to appreciate their contribution. Sometimes they reminded me of housewives aggrieved that no one noticed their labor while noisier people (such as politicians and managers) got all the attention. They took pride, bordering on self-righteousness, in the moral superiority of their work. Unlike managers and politicians, they valued objectivity. Over and over, they proclaimed that they understood the distinction between fact and opinion. Then—back to complaints—they lamented that engineers failed to leverage their disciplined ways of thinking into greater social influence. My grandfather especially regretted that most engineers limited their responsibility to the material dimension of their work. He wanted them to assume responsibility for the human dimension too. Engineers, he argued, should provide leadership to help working men and women achieve not only good pay and working conditions but also an understanding of their contribution to making a richer, fuller life possible for everyone.[8]

Even then I was aware of some ambiguities. On one occasion, when I was about twelve or thirteen, Grandpa, always the teacher, asked me what I thought of the profit motive. We were seated at the dinner table. When I hesitated (I did not have a clue as to what he was talking about), he pounded his fist down on the table so vigorously that the silverware jumped. "I *despise* the profit motive!" he bellowed. In Laurel, he went on to say, there were a few people who were motivated by monetary gain, but they were called misers or skinflints and were regarded with pity or contempt. The profit *system* was another thing entirely. As a relatively objective yardstick of the value of a contribution to the economic life of the community, it provided reliable guidance for future decisions.

Grandpa often made a similar distinction between the state and the community, which he expressed in somewhat mellower tones. The state could be beneficial and was certainly necessary, he explained, but the engineer's fundamental duty was to the community. Grandpa would linger lovingly on the middle syllable: com*mun*ity. Instead of offering a definition of 'community', he offered an illustration. While in graduate school in Breslau, a relatively large city, he had seen laborers digging ditches with picks and shovels. In the United States, in the same era, even a farm town like Laurel had "donkey engines" (small steam engines) to relieve people of such hard labor. This difference, he said, showed how engineers could serve the community.

Years later, when I read Edwin Layton's description of "the ideology of engineering" in his classic study *The Revolt of the Engineers*, I immediately recognized the themes of those sitting-room conversations. Layton notes that the "ideology of engineering" was in competition with the "ideology of business" in early-twentieth-century America. Engineers could achieve a "purely verbal reconciliation" between the two ideologies "only so long as engineers confined themselves to words."[9] The distinction between the profit motive and the profit system and that between the state and the community may have been "purely verbal reconciliations," but they served to maintain a strong if not entirely coherent sense of identity for engineers.

By identifying with the scientific method, engineers claimed a role as independent professionals, outside of social class, respected by workers and owners alike for their technical skills based on scientific knowledge. At the same time, most of them were employees in business bureaucracies, and they understood the value of doing for one dollar what any damn fool could do for two. Scientific judgment had to be weighed along with business judgment, and their identity as engineers was a constant effort to balance these two elements.

"The divided mind of the engineer," Layton writes, "reflects social as well as logical tensions."[10] The ambivalent relationship between engineers and management in twentieth-century engineering reflects their different social origins. Between 1880 and

1920—in the midst of the second industrial revolution—the number of practicing engineers in the United States multiplied twentyfold, from 7,000 to 136,000, and engineering enrollments in land grant colleges (including MIT) went up fortyfold.[11] During this period, engineering education allowed lower-middle-class or even lower-class young people (men almost exclusively) to get a professional degree in just four years. A 1924 survey showed that entering engineering students were "drawn from the poorer and less well-educated segments of the middle class." Many of their parents had small businesses or were farmers. Only 13 percent of their fathers had a college degree; only 40 percent had completed high school. Another telling statistic is that 90 percent of the freshmen in 1924 had to work a year before starting college.[12]

During the period of rapid professional growth that followed World War II, it continued to be the case that many engineers were from modest backgrounds. Even in the late 1960s, "a quarter of engineering graduates came from rural areas and another quarter from small towns."[13] Such students had a deep belief in meritocracy and in individual effort. Most of them did not identify with the managers, who tended to be from wealthier, better-educated, more metropolitan backgrounds. While accepting the basic rules of capitalism, most engineers maintained a disdainful view of managers who dealt in words ("management bullshit" was a common term) rather than with things and who did not really understand how things work. However, the contempt was mixed with envy. Managers were rewarded with higher salaries, more responsibilities, and more opportunities. It made sense for engineers to migrate into management, and many of them did. As early as 1933, the director of a major study of engineering education noted that by age 40 three-fifths of engineers were occupied with administrative rather than technical work.[14] In many cases the management responsibilities were functional (some aspect of a research, design, construction, or maintenance project) rather than general, but that did not make them less administrative. Even when engineering graduates migrated into administration, they often maintained a sense of professional identity, calling themselves engineers even when they were spending much of their time on managerial responsibilities.

The engineering profession, then, has had a complicated relationship with both science and business, but in the normal course of a career the tensions could be handled because they were sequential. In their educational preparation, engineers received heavy training in science and mathematics. Long before the postwar rise of "engineering science" it was understood that engineers needed grounding in these "fundamentals" to have command of basic principles that could be applied to a variety of technical problems. It was also understood that many if not all engineers would eventually migrate into managerial roles, where human and organizational work would dominate. The typical professional track began in science and ended in bureaucracy.

Modern engineering is an integrative profession, bringing scientific knowledge and business knowledge together for useful and profitable purposes. This integrative nature has been its glory, but now science and business have also become vigorous, integrative enterprises. Science gives rise to a stream of marketable innovations coming out of research. Business is dominated by the marketing of innovation, fed by scientific research and by capital investment. Engineering is less and less a separate realm and more and more an integral part of both science and business. In its relationship with science, engineering flourishes in developing devices needed in scientific research and in closing the gap between scientific discovery and marketable products.

## Engineering and Science

In a 1999 talk in MIT's Technology and Culture Forum, former provost John Deutch started by saying that he would talk about why he no longer believed in science and technology. This got the attention of the audience. He then presented a more "genteel" version of this provocative case by stating that the distinction between science and engineering was no longer very useful. "The world is now dominated by application, not by technology generation." In any project, Deutch said, there is "parallel processing" of thinking about the science, thinking about engineering issues like quality control, thinking about budget and marketing issues, and thinking about legal and political constraints. From the start, people work in groups that mix the range of disciplines necessary to get the project done. "We are faced," he said, "not with disciplines but with situations."

Though Deutch did not say so, his analysis implies the need to reconsider "engineering science," the dominant model for engineering research and teaching for the past fifty years. As early as the nineteenth century, engineers claimed the authority of scientific principles as a basis for their professional practice. My grandfather often said that MIT's chemical engineering department had pioneered engineering science in the early 1900s because chemical industries were evolving too swiftly for education and research to be based on current practice. Electrical engineers too were known for requiring their students to have a strong background in science (by which they meant math and physics).

After World War II, with strong leadership from MIT, engineering science became the dominant model. Engineering education was based on thorough grounding in scientific principles, and engineering research was based on models of scientific research. Federal funding supported the research model, since wartime experience had so vividly shown its importance to national defense. University hiring supported engineering science, which was highly compatible with the standard procedures for evaluating faculty members. Indeed, the rise of engineering science was critical in making engineering respectable as an academic discipline.[15] These factors combined to reorient engineering away from the shop floor and toward the scientific laboratory. Engineering never ceased to be an adjunct of industry, but it became much more an adjunct of science.

This reorientation has reshaped the epistemological relationship between science and engineering as it had developed in an industrial context. Members of MIT's engineering faculty had long assumed that industrial practice depended on students' understanding of science's "fundamentals": the laws of motion, the conservation of matter and the elements, the conservation of energy, the atomic structure of matter, electromagnetics, the principles of physical and chemical equilibria, the rules of chemical action, fluid dynamics, and the like. The idea was that science provided the relatively invariant and reliable substructure that underlay the more variable and malleable superstructure of industrial practice.[16] In the 1950s, Aero-Astro depicted this model as a tree with the basic sciences (math, physics, chemistry) as the deep roots, the applied sciences (fluid mechanics, thermodynamics, heat transfer) as the shallow root system, the engineering sciences (vehicle dynamics, flight control, guidance, structures) as branches, and vehicle system engineering at the top.[17]

Today the substructure-superstructure model is rapidly being displaced by the much more complex interactions of "technoscience," a convenient term that emerged in France in the 1980s to apply to the hybrid "situations" that Deutch described.[18] The word 'technoscience' defines a constant process of interaction in trans-disciplinary projects where the projects, not the disciplines, define the terms of engagement. In part this is because the science that is now most dynamic—biology—does not rest on first principles in the way that physics and chemistry do. More generally, the relationship between science and engineering is no longer summarized in a set of reliable equations; it now includes all the complexities of evolving life forms.

Participants in technoscience enter what the historian of science Peter Galison calls a "trading zone, an intermediate domain where procedures could be coordinated locally even where broader meanings clashed." In the trading zone of technoscience, engineers and physicists (to use the people on whom Galison focuses) invent ways of communicating through instruments and habits as well as through words—ways that Galison compares to pidgins and creoles—in order to coordinate a project even though they come at it from different professional cultures.[19]

The good news for engineers is that they are part of this robust zone. Traditional engineering disciplines are being brought into enormously productive interdisciplinary mixes that merge them in novel ways with one another and with various sciences (physical, biological, and informational). On the other hand, the zone is crowded. Everyone involved in the "parallel processing" of the project has to think integratively from the start. In these "situations," scientists may do much of the engineering themselves, and the separate identity of engineering, professionally and epistemologically, becomes less clear.

Still, the rise of technoscience does not mean that science and engineering are becoming one and the same. In science, the fundamental unit of accomplishment remains the discovery; in engineering, the fundamental unit of accomplishment is problem solving. But the whole point of a trading zone is that people find ways to trade despite different motivations and cultural contexts. Technoscience can flourish even when the motivations for and the definitions of accomplishment remain very different. How different they are emerged in the discussion after Deutch's talk, when some members of the audience commented on the different motivations that drive scientists and engineers. One graduate student in physics remarked that, while engineers might want to "solve the world's problems," he considered even the big problems of humanity to be, on some level, superficial. As a scientist, he said, he wanted to solve the big problems of nature, to make fundamental discoveries, to understand deeply how nature works—and he believed that ultimately its workings would be shown to obey simple, elegant rules.

Even if engineers are motivated by getting things to work rather than by understanding how nature works, they can collaborate successfully with scientists on projects. Nowhere is their success more evident than in the trading zone of biology and engineering. The MIT School of Engineering's new Biological Engineering Division is off to a roaring start, with 10 faculty members, strong connections to the School of Science and to local medical schools and hospitals, graduate degree programs, and a popular undergraduate minor in Biomedical Engineering.

Linda Griffith, the faculty member most influential in establishing the biomedical engineering minor, uses substructure-superstructure language in arguing that biology "must be considered as a foundation science of engineering, along with chemistry and physics." But Griffith goes on to emphasize that a foundational education in biology is not enough; biomedical engineers must remain in constant and strong collaboration with science: "The diversity of engineering itself prohibits a single cogent intellectual educational program of its applications in biology and medicine to be developed."[20]

In fact, one important rationale behind the creation of the Biological Engineering Division is to delay, if not to prevent, the emergence of a separate bioengineering department within the School of Engineering. Bob Brown (who served as MIT's dean of engineering before becoming provost) wants to keep what he calls "the two cultures" of biology and engineering in constant contact. In the age of technoscience, disciplinary and organizational boundaries, like national boundaries, can be a hindrance. Bioengineers in the new division have more in common with biological scientists, in many cases, than with other engineers in their home departments.

The Biological Engineering Division may become a separate department before long, but it would be a department with extremely porous boundaries and multiple linkages to other departments, labs, and centers. The vitality of trading zones, techno-scientific or otherwise, lies in breaking down boundaries, keeping things mixed up, developing a lot of interfaces, going with the flow. And the flow goes in both ways. Biological research generates a constant stream of projects that engage engineers, especially mechanical and chemical engineers. At the same time, biologists think more and more like engineers as they probe systems and mechanisms, worry about quality control, and construct large technical systems. The best of them often bring specialized technical knowledge to their biological research.[21]

A major factor in the success of this trading zone—and more generally a major factor in the rise of techno-scientific mixes—is the role of information technology in providing a common, readily transferable language: ". . . biology, electronics, and informatics seem to be converging and interacting in their applications, in their materials, and, more fundamentally, in their conceptual approach . . . the current process of technological transformation expands exponentially because of its ability to create an interface between technological fields through common digital language in which information is generated, stored, retrieved, processed, and transmitted."[22]

The interfacing effect of a "common digital language" is seen throughout MIT, not just in biology-related areas. All engineering departments are becoming, in some form or other, to a greater or lesser extent, departments of applied information technology. Accordingly, there is, in the words of an Aero-Astro faculty member, "a lot of intellectual convergence." Almost all engineering departments are trying to recruit what are charmingly called "computer science types"—adepts in the lingua franca of technoscience. Such "types" bring generalizing power to the department, but also tend to undermine departmental identity. It is hard to find someone who is expert in widely applicable computer-based systems and who is also interested in a particular industry or site associated with an engineering discipline. For example, the Department of Ocean Engineering is always looking for a first-class researcher in artificial intelligence, robotics, signaling, or systems, but it would like this person to have a special interest in the ocean. Some members of the current Ocean Engineering faculty do "wet" research; others do computer and systems work that is only incidentally related to salt water. Similarly, Aero-Astro has hired young faculty members with expertise in human movement and perception in technology-intensive environments who could just as well apply their research to non-aeronautical environments.

In the form of a common digital language, technology dissolves the familiar boundaries of engineering. It also lifts engineering, once the most down-to-earth of professions, from its familiar ground of materiality, endowing it with a new and ghostly lightness of being. Fewer and fewer faculty members in engineering actually make things or build things. More and more work with symbols and models. In a sense they are still working with machines, of course, but the meanings of both engineering and machinery are redefined when the machines process information rather than matter.

Again, to see the evidence one need only survey the internal redefinitions of MIT's engineering departments. In the early 1960s, Civil Engineering began to shift its research focus from structures to computing.[23] Today many members of the Civil Engineering faculty design not structures but software systems to manage construction.[24] As Civil Engineering has migrated from structures, Mechanical Engineering has migrated from machines. Many younger mechanical engineers are designing programs to replace machines, using electronics and computers to assume functions previously performed by more conventional mechanical systems. As one faculty member said: "Few mechanical systems are all mechanical anymore."

Similarly, Aero-Astro no longer identifies designing aircraft as its primary mission. In a recent strategic plan, the department proposes three "strategic thrusts": systems, information, and education. Compared to the materiality of structural mechanics or aerodynamics, these are all highly abstract areas of investigation, where symbols are manipulated as much as matter and where the role of information technology is highlighted.

The dematerialization of engineering is most evident, however, in the Department of Electrical Engineering and Computer Science, whose name proclaims the alliance of the material and symbolic worlds, the material properties of chip and machine design with the logical properties of computer code. The whole enterprise of "software engineering" reorients engineering from the manipulation of matter to the manipulation of symbols, or code. The hardware side of computer science is recognizably engineering, but it is not obvious where to draw the line between a software engineer and a programmer.

One side of engineering diffuses into the microscopic world of biology, another into the ethereal realm of cyberspace. Engineers always thought of themselves as doers; now some are logicians, sitting in an office all day, writing. Engineering is no longer necessarily applied science. It has developed its own theoretical wing, with practitioners who never actually build things and whose research takes them well beyond the range of common-sense experience. One junior faculty member says that he uses computer graphics as an "imagination amplifier." Another explains that he uses the computer not as a calculator but as a laboratory. The stuff of engineering turns into intangible experiments, symbolic objects, and abstract systems.

Centuries ago, science left the human scale behind to probe the unimaginably small and the unimaginably vast. Now engineers are pushing well beyond the limits of common sense: microscale heat transfer, picoseconds and femtoseconds, electrons to photons and back, molecular films, cell membranes, microscopic turbines. At one extreme, nanoengineering; at the other, global logistical systems. At one extreme, nanoseconds; at the other, an indefinitely sustainable earth. In the age of technoscience, science is not just a parent to engineering, but a lifelong partner.

## Engineering and Practice

The obvious direction for engineering, it would seem, is to build on this exciting and successful partnership. While this is happening, an equally strong and apparently contradictory trend in engineering is to move engineering education, and to some extent engineering research, back toward the realities of industrial practice.[25] Many of those who advocate that engineering get "back to practice" identify engineering science as the culprit that detached engineering from industry and the market. But engineering science is losing its dominance in part because of its own success: in mature areas of engineering, there is little fundamental science of interest left to explore. In aero-astro, structural mechanics and aerodynamics are well understood. In civil engineering, materials like concrete and steel are still essential for construction, but they offer dwindling interest for research. One older mechanical engineer concluded that the engineering science approach was producing diminishing returns: "Systems- and information-based approaches will likely yield bigger payoffs in many areas of engineering and manufacturing where the 'physics' has been well investigated."

In these situations, academically based engineers have two choices. Either they can migrate toward areas where the science is more exciting, or they can put new emphasis on non-science-based facets of engineering: how things are designed, how they are manufactured, how organizations work, how innovation is brought to market, and the like, all of which are often lumped together under the label "practice."

Another source of renewed interest in engineering practice is the renewed power of the marketplace in supporting engineering research. In the postwar period, governmental policies shielded huge sectors of engineering practice—defense contracting, highway construction, communications—from direct market pressures. Government patronage changed how academics thought about engineering: it made academics less tightly linked to industrial concerns and perspectives. Beginning around 1990, government support for research in science and engineering, both in corporations and in universities, began to level off in most areas (biology being the big exception). Since then, university-based engineers have sought to revitalize their links with industry as sources of research support and as employers of their graduates.

But industrial support for engineering research has been strongly market oriented. As government support began to flatten, industry began to take a more "value-received perspective" in regard to research investments: they had to be justified in terms of the bottom line, and the short-term bottom line at that. Some corporate labs become mission oriented; some disappeared. As investment in research diminished or scattered, consulting such as my grandfather did for Standard Oil became less important as a bridge between universities and industries. Instead, businesses found that they could get the benefit of good research ideas by investing in and eventually buying up small companies, which pay more attention to marketability, timeliness, and productivity than university labs. MIT found itself competing not only with other universities but also with startup companies—a fair number of which are connected with universities through their founders. The back-to-practice movement could also be called a back-to-the-market movement, in the sense of renewed and direct attention to profitability.

In the MIT context, there are two major ways that engineers define getting back to practice. One group of engineers looks to a renewed emphasis on design; the other group advocates a new emphasis on large technological systems. The interesting psychological and sociological differences between these two groups could be summarized as follows: the design advocates identify with entrepreneurs, the systems advocates with managers.

Some of the most outspoken design advocates are found in the Department of Aeronautics and Astronautics, whose head, Ed Crawley, unambiguously summarizes its three educational principles in this way: MIT educates engineers; MIT takes its educational cues from industry; and the institutional culture must change from that of engineering science to "CDIO." That punchy acronym, standing for "conceive, design, implement, operate," has become popular throughout the School of Engineering.[26] Other departments have developed variations on the theme; for example, Civil and Environmental Engineering has adopted the motto "plan, design, construct, operate" to express its departmental mission.

Many design advocates, whatever their age, have a sort of Young Turk quality. They challenge what they see as the engineering science establishment, asserting that they do *real* engineering by designing and building useful things that actually work. They may criticize engineering scientists, sometimes implicitly and sometimes openly, for dressing up ordinary insights with fancy but uninspired equations.

In contrast, advocates of the technological systems approach, who have congregated in the Engineering Systems Division, tend to be experienced individuals accustomed to working outside their departments or even outside their school (they most often collaborate with the Sloan School of Management). They are less brash than the Young Turks of design, partly because of their closeness to the managerial world and partly because their cause—seeing the world in all its complexity—encourages conciliatory attitudes. In particular, the ESD's advocates take a conciliatory view of engineering science. Most of them started their careers there, and they typically say they are trying not to overturn engineering science but to extend it to a more expansive definition of engineering.

According to its mission statement, the ESD "will establish engineering systems as a field of study that focuses on complex engineering systems and products, and views these systems and products in their broad social and industrial context."[27] Members of the ESD faculty emphasize the heterogeneous, integrative nature of engineering practice, in which technical design is always shaped by social, economic, and natural constraints. One participant explained that systems advocates are trying to show that "the -ilities matter." The -ilities here are scalability, flexibility, sustainability, maintainability, adaptability, reliability, and (admittedly lacking the suffix -ility) robustness and safety, all supposedly slighted by engineering science. In their lingo, they are trying to move from "little-e engineering" to "big-E Engineering."[28]

A primary goal of the ESD is revitalizing engineering education and reorienting it in the "big E" direction. It has been quite successful in that endeavor, in part because it had much to build on. Beginning in the 1980s, MIT pioneered educational and research programs that treat engineering in a socioeconomic framework of large-scale, complex organizations and projects. Many of these programs provide master's-level education for technically grounded senior managers. In particular, the Leaders for Manufacturing Program demonstrated MIT's ability to define a problem (in this case, the problem that American manufacturing was mired in non-competitiveness) and to make a significant contribution to solving it.

Leaders for Manufacturing and other educational initiatives continue to thrive. The vitality of some related research programs and the direction in which they are headed are less clear. The definition of a technological system remains vague. The concept generates lists of adjectives and nouns ('large', 'complex', 'dynamic', and the -ilities) rather than a well-defined set of problems. ESD faculty members have done research on topics as varied as improving the design of supply chains, overcoming adversarial relationships between labor and management, and improving corporate culture, as well as on issues of competitiveness and environmental sustainability. Different departments, however, tend to define systems according to implicit and dissimilar mental models. In Aero-Astro, the paradigmatic example of systems engineering is "putting a man on the moon." In Civil Engineering, it is a large construction project such as Boston's Central Artery/Tunnel Project (the "Big Dig"). In Mechanical Engineering, it is a manufacturing process, such as a "lean production" system for aircraft or automobiles.

It is difficult and perhaps impossible to define an "engineering system" in a way that covers these varied practices. In fact, once engineering is defined in terms of systems, practice may soon be left behind altogether. Systems are abstractions rather than things. Symbolic representations of complicated relationships, they are often expressed in mathematical form (utility functions, risk assessments, cost-benefit analyses). Systems analysts often come from a background in mathematics or sciences rather than in one of the traditional engineering disciplines. Some of them try to develop mathematical representations of complex systems that will constitute a sort of Grand Unified Theory of Engineering.

In a similar though less pronounced way, design advocates too may be seduced by the possibility of developing an abstract, generalizable science of design. For example, Neil Gershenfeld of the Media Lab is working on an organic "new theory of design . . . that is not hardware or software." This "engineering theory of life" would manage complexity by "specifying processes that result in the emergence of a working system rather than by specifying the design of the system itself."[29] Engineering design and engineering systems, then, are studied in two very different ways: while some advocates emphasize the need for intuition and awareness of messy complexity, others seek rigor, abstraction, and calculation.

With the line between science and engineering becoming so blurred, and with design advocates and systems advocates split between everyday practice and grand theory, what is real engineering today? This question can lead to some quasi-theological debates, but at a university it must be asked for a highly practical reason: so that wise decisions can be made about hiring and promoting faculty members. One strength of MIT is that most promotion cases and all tenure cases are heard by all the deans and other academic members of Academic Council, which is chaired by the president. For example, the Dean of Engineering listens to and votes on history cases, and a historian who serves as Dean of Students and Undergraduate Education listens to and votes on engineering cases.

For nearly every engineering case, those of us on Academic Council would start with an exercise in classification. What kind of a case is this? What type of engineering does this individual do? To answer, we would situate the case in three dimensions. First, was it an engineering science case, a design case, or a systems case? Second, was it cutting-edge engineering (where exciting research is going on) or maintenance engineering (something that ought to be taught, but not a lively research field)? Third, was it a type of engineering that belonged at MIT, in view of MIT's institutional mission, or was it better suited to another kind of educational institution, or to industry, or to another part of the world? Only then would we concentrate on the merits of the individual candidate. As a rule, the closer an engineering case is to science (whether engineering science or technoscience), the easier it is to review. The normal academic yardsticks fit: quantity and quality of publications, quality of academic peer reviews, comparisons with other scholars. In contrast, design cases and systems cases are usually knotty. The strong point of design cases is that, in one participant's words, "at least they make things that work." But that is also the weak point: in the words of another participant, "we can't just populate the place with people who build things." An individual who has already established a solid record in engineering science can move with credibility into design or systems research. To begin there, however, is to take a risk—not necessarily huge, certainly not insurmountable, but still a risk.

In an academic setting, an engineer who seeks tenure as a "design case" may have difficulty establishing a publication record. There are many places to publish concepts, but there are not many places where one can publish designs. It is possible to provide a list of inventions or innovations rather than a list of publications, but it is not clear how to compare a product or a device with a scholarly paper. Also, it is difficult to apply the traditional scholarly standard of "impact on the field" to design. It is one thing to design a prototype that can be demonstrated; it is another to bring it to full-scale production in a way that might have significant social influence. And who should evaluate impact? Who are the peers of a designer? When recommendations come from a mix of academics and non-academics, it can be difficult to interpret the scholarly standards of the latter.

Academic Council faced similar problems with engineering systems cases. There were fewer of them, because systems people tended to be internal émigrés rather than new hires. The cases we did consider were at least as hard as design cases. For systems people, the unit of accomplishment is often solving problems that are essentially managerial: shutting down a factory, dealing with public disputes, directing complicated salvage operations. In the real world, these are important problems. For the academy, what is often missing from the case is the element of reflection on the problem. To some extent, our promotion and tenure discussions are simply reminders of the imperfect fit between academic and real-world contexts. That there is a gap is not the problem. There should be a gap between academic engineering and the larger world: the university should provide critical reflection on the larger world as well as service to it.

The difficulty, for Academic Council, lay in deciding the epistemological status of building things or dealing with managerial problems: in other words, the intellectual status of doing or making as opposed to thinking. Implicitly, and sometimes explicitly, we were placing candidates on a scale of mental activities, from routine ones to ones demanding intelligence to ones requiring brilliance. In these discussions, I found myself recalling the efforts of Enlightenment thinkers to distinguish the varieties of intellectual work performed by artisans, artists, and savants or to develop a scale of accomplishment ranging from that of the mechanic to that of the genius. Like the *philosophes*, those of us on Academic Council sometimes experimented with new terminology to make such distinctions. In one case, a candidate for promotion was defined as a "technologist." When I asked how a technologist differs from an engineer, I was told that an engineer looks more at the science, while a technologist only builds devices. We even invented a

new category of professor—the "professor of the practice"—
to provide a suitable academic niche for individuals whose
accomplishments were more practical than scholarly.

The promotion discussions on Academic Council reveal the
extent to which technology has outgrown engineering. The
category of "engineer" is far too limited to cover the ways
individuals now engage with the technological world, not to
mention the many more ways that they will engage with it in
the future. Both the design movement and the systems engineering movement seek to reclaim a distinctive identity for
engineering—to proclaim that here is something engineers do
that scientists and businessmen do not do. In the end, however, the reclamation efforts only underscore engineering's loss
of identity. In both design and systems work, many people
other than engineers are in on the act. In design cases, those
of us on Academic Council would often end up comparing
engineer-designers with architect-designers and experimental
scientist-designers. Particularly in the language-based, highly
socialized realm of information technology, design involves
the logic and aesthetics of conceptual structures as much as it
involves the design of the materials that make up chips, servers,
and cables. In design today, engineering, programming, science,
language, and art converge.

In dealing with technological systems, it is even more obvious that engineers have to collaborate with political scientists,
economists, lawyers, and managers, just for starters. At MIT
the constant dilemma is whether to hire these collaborators as
ESD faculty members or to try to get other departments and
schools to hire them. On the one hand, the leaders of ESD
want, as one of them put it, "someone who understands our
world." At MIT, furthermore, researchers may find it an
advantage to be associated with engineering because it remains

at the core of MIT's institutional identity and value system. One social scientist considering an ESD appointment admitted that he preferred to be placed there rather than in a non-engineering department because "it gives me a certain legitimacy here [at MIT]." Outside MIT, however, the situation may be reversed. A political scientist or a legal scholar who takes an appointment in the School of Engineering may not be well placed to progress, much less excel, in his or her own discipline. As a so-called engineering school, MIT's standards of legitimacy are unusually well defined but also unusually limited. "Someone who understands our world"—meaning MIT's world—may not necessarily excel at understanding the larger world.

Meanwhile, the larger world does not just sit there outside the walls of the university. The world is always entering in the form of new students, who turn over much more rapidly than do faculty members. MIT students today are just as intent as any member of the engineering faculty in getting engineering closer to practice, in the sense of more meaningful engagement with real-world problems. Most students do not think that either design or systems is the way to do this, however. They have their own ideas of what engineering practice is all about. In acting on those ideas, they are at least as powerful as Academic Council in reorienting engineering education at MIT today.

## A New Breed of Engineering Student

In 1989, Robert Weatherall, then the director of MIT's Office of Career Services and Pre-Professional Advising, reported that he was seeing "a new breed of engineering student" characterized by greater social sophistication, better communication skills, and more desire for self-direction. More students came from upper-middle-class professional families. Fewer were small-town boys, far fewer were farm boys, and fewer were boys—women continued to flood into colleges and the work force. Increasing competition for college allowed engineering schools to be more selective. In 1982, for the first time, the average combined SAT scores of entering engineering students surpassed those of all non-science/math majors.[30] By 1985, the National Academy of Engineering reported a consensus that these new students "have a richer educational and cultural background and are more confident, more assertive than engineering students of years past."[31]

One important element in this emergence of a "new breed of engineering student" was the much wider variation in ethnic and geographical background. If you look at MIT yearbooks from the past, though you may not see into the souls of the students, you can certainly see the external homogeneity of the student body. In the 1949 yearbook, for example, out of 505 student photographs (the entire class numbered 900), you see one Indian, one black, and seven women. (The 305 faculty members pictured include one black and no women.) The contrast with the present could not be more startling. In particular, the number of Asian-American students quintupled between 1979 and 1989. Students of Asian heritage now make up roughly one-third of the undergraduate student body.

But this is only one element of the amazing diversity at MIT, which can be seen every day in the Institute's corridors. For me the most striking indicator of the diversity of engineering students today is linguistic: one-third of MIT's undergraduate students speak a language other than English at home. Each year as dean I set aside large blocks of time before commencement so I could practice pronouncing the names of the graduates, often using phonetic spellings or recorded voice-mails left by the students so I could get them more or less right.

The internationalization of the student body also meant its urbanization. Many more of these students were from cities, fewer from farms or rural areas. Once the shift was underway, it fed on itself as more diverse and sophisticated students were attracted by more diverse and sophisticated students. These students were not necessarily engaged in MIT's culture of "nerd pride." They saw engineering less as an end in itself than as a route to a business-oriented career. Many of the "new breed" did not share the anti-managerial prejudices of pre-World War II students from the American lower middle class. Instead, they were unapologetically interested in business and administration from the start. Weatherall notes that even the student of the 1980s was more eager for responsibility and did not want to "slog his way through a series of narrowly focused engineering jobs before he counts for something." He, and increasingly she, had different expectations. As early as 1986, when asked what they were looking for in their careers, 42 percent of the students graduating from MIT with bachelor's degrees rated it as essential or very important that they have an opportunity to become a top manager, and 28 percent of them were intent on having their own businesses. In all, 53 percent of the bachelor's degree recipients checked one or the other as essential or very important.[32]

During the 1990s, the trend toward hybridizing engineering and management only became more pronounced. In the summer of 1999, a survey by McKinsey & Company of post-1971 engineering graduates with bachelor's degrees showed that two-thirds of the MIT respondents were working in startup firms or in startup divisions of larger companies. Forty-two percent were working in companies under five years old, and

75 percent of the MIT respondents of the 1990s expected to leave their first job within three years. Thirty percent of the MIT respondents from the 1990s aspired to become entrepreneurs; only 21 percent aspired to be "technical leaders" (defined as senior scientists, lead designers, and research fellows).

For all engineering respondents, including those from MIT and other leading engineering institutions, the numbers were similar: 24 percent wanted to be entrepreneurs and 29 percent technical leaders.[33] The engineer used to stand in an indirect relationship to the market, typically acting through a business that someone else owned. Now many engineers intend to be owners too. Even for engineering graduates who have no entrepreneurial ambitions or formal managerial responsibilities, involvement with business activities, as opposed to technical activities, begins early. Among engineering graduates from the 1980s, even those who have no or few formal managerial responsibilities report that they spend significant time in sales, marketing, purchasing, and employee relations.[34]

In the words of one faculty member in MIT's Department of Mechanical Engineering, engineering and management are the "hot mix." Once again, information technology acts as a solvent. Engineers with an IT background can enter the market directly as entrepreneurs, or they can apply their skills directly to financial and management consulting. Weatherall distinctly remembers the moment in the 1980s when a recruiter from the financial firm J. P. Morgan first appeared in the Career Services office, interested in interviewing MIT seniors. This hint of something blowing in the wind was followed by a gale of recruiting activity on the part of financial and management consulting firms. In 1994–95, 25 percent of the companies recruiting at MIT were consulting or financial firms, and another 19 percent were software firms. Together, then, these companies made up close to half of the recruiters, up from 15.2 percent in 1982–83. This trend held firm through the year 2000.[35] In 1994–95, only 4.7 percent of the recruiters represented companies in traditional engineering fields (oil, chemicals, materials, food), and only 7.7 percent represented defense or energy contractors. The largest area of recruiting in manufacturing was computers, communications, and other civilian electronics (24.9 percent).[36]

In Academic Council, when we were discussing declining enrollments in Aero-Astro and other industry-oriented departments, someone asked why students are less interested in such departments today. Someone else cracked: "They read the front page of the *Wall Street Journal*." In another conversation, a member of the MIT Corporation summarized the situation this way: "In engineering too the money is at the intersection of technology and the market." This is the crossroads where MIT students are headed. MIT students usually declare their major at the beginning of the sophomore year. Among the 984 who became sophomores in the fall of 2001, 317 declared electrical engineering and computer science as their major; 112 declared management science, which for the first time became MIT's second most popular major, far outstripping biology (79), mechanical engineering (74), and chemical engineering. It could be argued, then, that none of the top three majors of the class of 2005—EECS (where computer science dominates), management, biology—is an engineering major in the traditional sense.

Whether or not they succeed as entrepreneurs or as managerial consultants, today's students have already changed engineering by heading for the crossroads of technology and the market rather than for a rung on the organizational ladder. One MIT faculty member told me: "It seems to me that we have long ago given up training our students as 'Professional Engineers' in most departments. We are really training technology innovators. . . ." This does not mean that engineers have suddenly lost their role as "bureaucrats" (Layton's term). Many engineers will continue to be employees, and even the self-employed are still working in organizational contexts. The context is different, though, because industrial organizations have changed dramatically in the past generation. They are now constructed to maximize flexibility in every way possible: through flatter, decentralized hierarchies, through corporate strategic alliances, through "just-in-time" labor and supply chains. Engineers and other technically skilled employees are part of this flexible, impermanent world. They are highly valued because of their skills, but their careers are highly unpredictable.

No longer is the corporation the rock on which an engineer builds a career. Now the corporation is more like a stream, a conduit for the flow of innovation. The individual engineer must construct a career in the midst of this stream. Engineers, now much more on their own, need to focus on staying with the flow, not with the organization. They have to be able to work with teams on an endless series of "situations," not always with support from a company, not commonly shielded by reliable government contracts, exposed to the relentless pressures of the market, constantly having to market themselves. Engineers will always be "organization people," but now they are more active in choosing their organizational contexts. For better or worse, students who go into engineering understand that their role will be shaped less by external organizations than by their own internalized drives as they engage in a continuous process of self-promotion.

The identity crisis of the engineering profession arises from this institutional transformation. Engineers now depend on themselves to build their professional identity in an organizational landscape dominated by networks rather than by clearly bounded corporate units. When engineers worked primarily for relatively stable mid-size or large corporations, their identity was defined in terms of contribution to those organizations. The ideology of engineering assumed that the engineer's loyalty to the organization would be rewarded by the organization's loyalty to the engineer. That social contract has long since been cancelled, as businesses have repeatedly cut their engineering staffs to protect the bottom line. Engineers have learned that they are part of the market flow rather than the corporate structure. In many ways this is liberating. They are freed from dependence on one company, and often an individual can have a much more interesting career. But market-based engineering also demands relentless hard work and entails exposure to risks of financial uncertainty, especially as an individual ages.[37]

Students understand better than many faculty members that this organizational context has redefined the meaning of engineering practice. Students pragmatically shape their educations and their careers to fit current practice. They are more interested in preparing for a technically oriented career than in preparing for an engineering career. In a hybrid world, technically oriented workers, whether self-employed or employed by others, can no longer define problem solving in material terms alone. Their careers demand a mix of technical, human, and organizational skills. In the global economy with ever-shifting, loosely bounded institutional structures, they need a broad education so they can navigate the seas of change more or less on their own.

This broader education is a "liberal" one, not in the classical definition of an education free from utilitarian goals, but in the practical sense of an education that provides many career choices over a lifetime in an uncertain world. Engineering education remains functional and instrumental. The difference is that now functionality and instrumentality include many organizational and human skills along with technical ones. In this limited but not trivial sense, engineering education is being humanized.

## A Broader Educational Mission

For most of its history, MIT has had an institutional commitment to a broad engineering education, but its success in accomplishing this has fallen short of its rhetorical commitment. Now economic realities are finally doing what MIT as an institution has long struggled to do for loftier reasons: broaden engineering education.

From its nineteenth-century origins, the Institute expected its students to have "exposure" (the conveniently vague common term) to economics, to history, to the social and ethical principles of organizations, and to communication skills, especially writing. This "humanistic stem," as it was called in engineering education circles, was not intended to give a gentlemanly gloss to engineering students; this was not central to their educational or professional expectations. Instead, the "humanistic stem" was intended to provide education in skills that were clearly professional but were not strictly technical. The reasoning was that engineers would outgrow their technical training in relatively short order, and furthermore that their scope of activity and influence would be severely limited if they had only technical training. Success and fulfillment over a professional lifetime required that a technical education be supplemented by a more enduring general education.

This reasoning was sensible then and is just as sensible now. Putting it into practice has always been a challenge, however. For most of MIT's history, the "stem" was given relatively low priority by faculty members and students alike, while the educational "tree" was and has remained solidly technical and scientific. In an institutional setting focused on science and engineering, faculty members who organized the curriculum never really had to defend their priorities against competing priorities. Before the "new breed" started coming along, students did not typically enjoy studying humanities and social sciences, nor, with their focused career plans, did they commonly see the point of such study. Finally, corporations tended to see engineers as employees—loyal and hard-working ones, they hoped—rather than as potential leaders.

In the years after World War II, however, non-technical education began to get more attention, for reasons that went far beyond any practical needs of the workplace. Wartime experience had demonstrated that the devices, systems, and decisions of engineers and scientists affected not just industry, not just the economy, but the course of civilization. Faced with this sobering fact, MIT's leaders decided that engineering students should be educated in a way that was commensurate with their social responsibilities.

On August 5, 1946, almost precisely a year after the atomic bomb was dropped on Hiroshima, James R. Killian, then an MIT vice president, wrote a two-page memorandum to President Karl Taylor Compton proposing that Compton appoint a faculty committee to study educational objectives, organization, and operations at MIT. Killian provided a list of some "very fundamental questions" for the committee to consider. For example:

Have we swung too far in the direction of theory as distinct from practice, or have we not gone far enough? Have we yet allotted enough time for humanistic studies? Do we require too much routine and scheduled work of our students, leaving too little time for self-development?

Killian suggested that "W. K. Lewis [was] the logical person to be chairman because of his great prestige and his strong interest in teaching," although it would be difficult to persuade him because of his many other responsibilities.[38] Compton did persuade Lewis, however, and the full committee, which included Elting Morison and Gordon Brown, began its work in 1947.

The report of the Lewis Committee (as it is called to this day) was presented to the faculty in December 1949. As its central recommendation, the committee called upon MIT to assume "A Broader Educational Mission." That, indeed, is the title of the decisive third chapter of the report, which begins by summarizing the prewar belief in engineers as capable of general social leadership—the same belief Grandpa expressed so often in the Newton sitting room, but in more elevated language:

We recognize especially a need to develop a broader type of professional training that will fit engineers to assume places of leadership in modern society, and we believe that this can be accomplished in part by improvement in the professional subjects themselves, and in part through further strengthening of the natural sciences, the social sciences, and the humanities. . . .[39]

The third chapter goes on to sound a new note, reflecting wartime experience, in describing how a changed world has changed the mission of engineering. The report does not use the words 'reflexive' and 'hybrid', but it expresses the concepts clearly:

In our increasingly complex society, science and technology can no longer be segregated from their human and social consequences. The most difficult and complicated problems confronting our generation are in the field of the humanities and social sciences; since they have resulted in large measure from the impact of science and technology upon society, they have an intimate relationship with the other aspects of the MIT program.[40]

The report called for the creation of a school of humanities and social studies at MIT. There were already programs in these fields, but the committee felt that the time had come to unite them in a school having equal standing with the schools of architecture, engineering, and science. The School of Humanities and Social Science was established in 1950, the same year James Killian was inaugurated as MIT's president.

Along with many MIT colleagues, including many in science and engineering, I still find the Lewis Report, especially its third chapter, prophetic and soul-stirring. Its practical results are also impressive. The School of Humanities, Arts, and Social Sciences (as it was renamed in 2000, on the occasion of its fiftieth anniversary) flourishes as a center of scholarship. It plays an important role in undergraduate teaching: under the General Institute Requirements, every student takes the equivalent of one class each term in humanities, arts, or social sciences. Term after term, year after year, these classes provide valuable "broadening." The role of the humanities, the arts, and the social sciences at MIT is distinguished and significant, but it does not accomplish the broader, socially responsible engineering education envisioned by the Lewis Report. Students are usually grateful for their classes in the humanities, the arts, and the social sciences, but the good effects of those classes are often serendipitous. For many (probably most) students, these classes are of secondary interest, and in any case they are perceived to be largely detached from scientific and engineering studies. Very few students major in the School of Humanities, Arts, and Social Sciences (with the important exception of economics, which is taught in a highly mathematical style). The school offers some first-rate graduate programs, mainly in the social sciences (again economics stands out). It offers only two doctoral degrees in areas close to traditional humanistic disciplines: one in philosophy and one in the history and social study of science and technology.[41]

The good news is that engineering education is becoming more and more socialized. The bad news is that this is happening almost entirely under the aegis of business. Whatever the official curricular requirements, many students regard economics or management as a de facto requirement. In a 1999 poll, half of the undergraduates said they would minor in management science if such a minor were available. Since it is not, many students major or double-major in management science. Students are looking for an education that is socially aware in a very pragmatic sense. They are increasingly aware that they need to know how society works, which is not necessarily the same thing as an education that highlights social responsibilities.

As a result, engineering education today is, as we say in the humanities, contested terrain—a site where different strategic goals collide. Most students want a socially aware (in the practical sense) and technically oriented education, which is not a traditional engineering education. Design advocates and systems advocates want to reorient engineering education away from engineering science, but no one really wants to give up the strengths of an engineering science education. Yet the boundaries between engineering and science keep getting fuzzier as the biological sciences challenge the whole idea of "fundamentals" and MIT continues (though not always vigorously) to back the concept of a truly broader engineering education, for hovering over all this is the ghost of the Lewis Report: the realization that we live in a world where the consequences of science and technology have long since broken through the walls of industry and of the laboratory to become an integral and determining element of nature and history.

All these habits and convictions and purposes are thrown into the curriculum. The result is a logjam. Everyone concerned with engineering education yearns for dynamite, if only they could agree on where to set the charge. From the faculty's point of view, designing an engineering curriculum is one of the most difficult design problems in engineering today. One faculty member joked that he truly wondered if degree requirements at MIT were a computable function. Faculty members may joke, but it is the student who has to cope with the multiplicity of uncoordinated demands. For today's engineering students, managing an undergraduate schedule is the most challenging of all the challenges in engineering that involve the management of large, complex systems with multiple constraints and multiple goals.

What engineers are being asked to learn keeps expanding along with the scope and complexity of the hybrid world. Engineering has evolved into an open-ended Profession of Everything in a world where technology shades into society, into science, into art, and into management, with no strong institutions to define an overarching mission. All the forces that are pulling engineering in different directions—toward science, toward the market, toward design, toward systems, toward socialization—add logs to the curricular jam. The most obvious symptom of the identity crisis of engineering is the crisis in engineering education.

## The Total Curriculum

In the post-World War II years, MIT, taking its cue from the Lewis Report, erected an elaborate curricular structure as a sort of "great compromise" intended to balance the competing claims of science, engineering, and humanities as elements of a technical education.[42] During the past several decades, new demands, both faculty-driven and student-driven, have both drilled into and added onto this edifice. It still stands, but the grand facade is nearly obscured by all the modifications.

Take, for example, the quarter of the curriculum that the "great compromise" devoted to the "science core." The rise of technoscience and of bioengineering begins to dissolve the concept of a well-defined "science core," in the sense of indisputable laws that furnish a stable foundation for all forms of engineering practice. Instead, engineers are in constant interaction with science, which is also constantly evolving. What is fundamental to one engineer seems irrelevant to another.

The event that signaled the breakdown of the old substructure-and-superstructure justification for the science core came in 1992 when the MIT faculty added biology to the science core requirements (along with existing requirements in physics, math, and chemistry). By almost any measure, the addition of biology has been a great success. As the popularity of the bioengineering minor indicates, many undergraduates use the subject as a foundation for engineering studies. For others, including many students majoring in electrical engineering and computer science, the biology requirement is not perceived as foundational. These students enjoy the biology subject and benefit from it, but for them it is a broadening and not a foundational experience. Conversely, many students who are going into biology or bioengineering regard physics as broadening rather than foundational. The faculty members who teach freshman physics are well aware of this shift in expectations. At one meeting with a small group of math and physics faculty members, I was involved in a discussion about new teaching methods for the freshman year. The faculty members all agreed that, although MIT students understand the importance of math, few of them see physics as relevant. "At MIT now, physics is like physics for poets," one member of the physics faculty remarked. I was shocked: MIT has always prided itself on not "watering down" the science core. "Do you really mean that?" I asked. My colleague explained that the subject is by no means watered down—in fact, freshman physics is notoriously difficult—but that the students look on it as a mind-stretching experience rather than a functionally useful one.[43]

Now that a new subject has been added to the core, and now that the purpose of old subjects is mutating, the whole logic of the core has begun to weaken. For example, why not add other mind-stretching sciences to the core? Some members of the Earth, Atmospheric, and Planetary Sciences faculty contend that the core should include environmental science. Besides having great practical importance, they argue, it represents an alternative approach to scientific problems (especially in its use of modeling techniques) that students should understand and be able to use. Why not insist that students get both ultra-descriptive sciences and mathematical sciences? These arguments are hard to refute. The objection to them is not logical but practical. Already, with increasing pressure from science on one side and from humanities, arts, social sciences, and management on the other, engineering departments find their share of the curriculum in a vise at a time when they have more and more that they want to teach. It is, roughly speaking, the curricular equivalent of Moore's Law: every 18 months the amount of information one would like to cram into the head of a student doubles.

The pressures to expand engineering curricula do not come from the sheer increase in the volume of information, however. Departments are also trying to educate students in sifting, sorting, and handling information, which they often do through integrative "project labs." In John Deutch's terminology, project labs give students experience in dealing with "situations" where learning involves not just solving a problem but also defining the problem in the first place.

Project labs also serve as recruiting tools for engineering departments. Students usually like them. Many students came to MIT in the first place because they like building things and being with others who like to build things. They can find it something of a shock when they confront a curriculum that transforms familiar, "fun" technologies into unfamiliar abstractions. Students easily lose their enthusiasm when confronted with a curriculum that is heavy on problem sets and light on face-to-face interactions with other students and instructors. Problem sets make a sturdy rudder, but education also needs a breeze to fill the sails. Engineering departments find that team-based project labs help provide a motivational breeze, attracting students to majors and keeping them there. Though the abstract principles of engineering may be the same across all its varieties, the "stuff" of the disciplines is typically what motivates students. Aero-Astro students want to work on space travel, mechanical engineers want to build nifty robots, chemical engineers want to design medical processes, and so forth.

The need to attract and hold students is more and more of a challenge for engineering departments at MIT because of the overwhelming popularity of electrical engineering and computer science as a major. EECS attracts half of all engineering majors, or one-third of all undergraduates. Many students who major in EECS do so for the simple reason that they have grown up loving computers and programming. The fact that past EECS majors have had their pick of well-paying jobs has also been a big incentive. For students who are at all undecided about a major, EECS is often the default mode. Whatever they want to do, they know they will end up with an educational platform that affords them maximum professional flexibility in the information age.

EECS is such a popular department at MIT because it fills a dual educational function. It is an undergraduate major in the traditional sense for students who intend to become electrical engineers or computer scientists. In the age of information technology, it is also a general degree, filling the role that mechanical engineering used to play for engineers or the role that English used to play for liberal arts majors. Because computer science fills this role of general education, many students act is if it were a requirement, though officially it is not. Undergraduates pile into the five freshman classes in basic computer science, which, tellingly, are offered by five different engineering departments. After the first year, they look to their majors (even if they are not EECS majors) to provide more specialized education in computer skills. This puts yet another source of curricular pressure on engineering departments: to add more computer-related education to their major curricula. Almost all of them do this through use of computer-aided design tools (such as matlab and statview). Some offer an even more computer-intensive curriculum, so that they can publicize their major as a form of applied information technology.

For example, Aero-Astro recently initiated a new undergraduate degree that it calls a bachelor of science degree in aeronautics and astronautics "with information technology." When the proposal was discussed in faculty meeting, an engineer from another department noted that almost any department at MIT could now, or very soon, offer such a "with IT" degree. Someone else at the faculty meeting commented that in twenty years we would not imagine having such a designation added to the degree: we would take it for granted. The Aero-Astro faculty member presenting the proposal did not argue the point, saying only that it was important now for students to have this notation for the world to see. The measure was approved, since everyone at MIT—EECS faculty most of all—want to see students spread out from EECS to other departments.

Efforts to attract students to other branches of engineering cause a dilemma, however. Students who are not intrinsically fascinated by the concrete stuff of an engineering discipline—space travel, robots, molecules—will probably keep heading for the hard-core curriculum of EECS rather than for a more constrained version of information technology. On the other hand, students who are fascinated by the concrete stuff are not necessarily going to be attracted by the abstract language of information systems.

The larger challenge for engineering departments is to try to do everything they feel they must do—retain a deep commitment to engineering science, add more practice-oriented labs, keep up with the information explosion in the field, provide sufficient education in information technology—in the traditional four years of an engineering curriculum. Because there is not really time for all this in four years, students display great dexterity in curricular multi-tasking. Many overload, taking an awesome (as they would say) number of credits each term; many double major, so that they end up with two bachelor's degrees. One reason for doing this is cultural display (MIT students take pride in showing off their academic intensity), but another reason is practical. For example, since students cannot minor in management science, many of them declare it as a second major instead. The piling up of majors was getting so out of hand that a faculty committee recently ruled that a student could declare no more than two majors.

Another prime example of academic multi-tasking is MIT's communications requirement, which took effect with the incoming class in the fall of 2001. From the start of discussions about reforming an earlier, less rigorous writing requirement, MIT faculty members took it as axiomatic that they could not add any new free-standing requirement. This effectively ruled out the option of a "freshman comp" class, though, as it turned out, there were sound educational reasons for choosing an alternative approach. The agreed-upon solution involves integrating instruction in writing and speaking skills into classes that students have to take anyway, both in the humanities and social sciences and in engineering and science. It is a creative solution, and it is likely to be an effective one. The inability even to consider other options, however, indicates how bad the curricular logjam is.

The crisis of engineering education is evident far beyond MIT. For many years, the Accreditation Board of Engineering and Technology, a nationally powerful organization familiarly known as ABET, has provided general regulation of engineering curricula. Its accreditation procedures have long specified, within certain parameters, the units that departments should teach in various areas. For the most part, MIT's engineering departments have taken these criteria seriously, so most students receive ABET-accredited degrees. But in the late 1990s ABET threw in the towel, or, to use a more appropriate metaphor, punted the ball. It gave up trying to specify the subjects that engineering departments should teach. Instead of this "input-based" approach, ABET asked engineering departments to use an "output-based" approach, developing their own ways of attaining curricular goals and figuring out for themselves how to fit it all into the standard undergraduate program.

As ABET no doubt understands, handing the problem over to engineering institutions is not solving it. The lid will keep popping off engineering curricula as long as the expectation of a four-year degree stays in place. The four-year limit has been an article of faith among engineering educators for a century, largely as a result of a commendable desire to keep engineering a career open to talent. It provides upward mobility for young people who could not afford undergraduate general education followed by more years of professional education. At the same time, the four-year limit has frustrated efforts to raise the professional stature of engineering when other professions insist on a more extended and expansive education. The Lewis Report reaffirmed the four-year standard, although that recommendation undermines its call for "a broader educational mission." My grandfather privately felt that it was time to extend engineering education beyond four years: he felt that broadening ultimately requires lengthening. However, he also understood the practical pressures on an institution like MIT to keep engineering education relatively short and inexpensive, especially compared to other forms of professional education.

That was in 1949. Today, engineering educators realize that moving the goal line may be preferable to punting. The financial pressures are still there, but more students understand the importance of investing in a higher-quality education, especially when they can earn enough to pay off any debts in relatively short order. Accordingly, MIT has begun to extend engineering education. The prototype of a longer degree is the five-year Master's of Engineering program introduced by the Department of Electrical Engineering and Computer Science in 1994 and now awarded by a number of other engineering departments. In EECS, students who complete an integrated five-year program obtain both a master's and a bachelor's degree from the department. The four-year bachelor's degree is still available, but the department is quite explicit in proclaiming the five-year program as a "first professional degree." In translation, this means roughly "Anything less is not really enough." The Master's of Engineering program was popular from the start: only a department of such stature could have succeeded in attracting so many students to invest in another year of education when the job market was so robust.

Adding a year to engineering education does not necessarily reform its character, however: longer is not necessarily broader. Most five-year programs at MIT, including the EECS prototype, focus almost entirely on technical skills. The character of the M.Eng. degree is evolving as the EECS department moves toward including in the requirements a "broadening" subject based on historical case studies of engineering innovations. As MIT slowly changes its curriculum, more and more MIT students are simply rejecting the four-year degree limit by choosing to go on to some form of graduate education. In the spring of 2000, when the lure of dot-com success was at its peak, 43 percent of MIT's engineering graduates went to graduate school, only slightly lower than the number who went to full-time jobs (47 percent). The percentage is even higher among those who have majored in science: 58 percent of them went to graduate school and 25 percent to full-time jobs.[44] One out of five MIT undergraduates is a pre-med; the percentage who actually apply to and enter medical school is lower, but interest in medical careers shapes their undergraduate educational experience. As time passes after graduation, these percentages rise. Twenty years out of MIT, 80 percent of MIT students have earned graduate degrees. More and more often, students major in engineering and sometimes even do graduate work in the field with no intention of practicing it. The tacit assumption of many if not most students is that they are getting an engineering education that will allow them to do any number of other interesting, useful, and well-rewarded things. Those "things" will probably have a technical orientation, but this is a very different statement from saying that the individuals in question will be engineers.

The real issue here is not the future of engineering education but the future of higher education. At an international meeting in Zurich devoted to the future of engineering education, one participant asked, only half in jest: "Why is it that we care so much about these narrow-minded engineers?" He answered his own question: "Because they can do more damage than most people." I would modify that answer: We care so much about engineering education today because so many people other than engineers are directly concerned with technology and can do a lot of damage. We care because engineering education must be reintegrated with general education—and what we really need to care about is general education for a world that is both technological and democratic.

### Engineering in the Common World

In the late 1930s, Grandpa Lewis and his students worked on using catalysts to help crack petroleum (that is, break up the big molecules into smaller ones to yield higher-octane fuel). In a small basement lab, they worked for months with an experimental apparatus, trying to keep the powdered catalyst well mixed with a stream of petroleum vapor as it moved through the apparatus in order to maximize the catalytic reaction.

The catalyst kept slipping out and settling to the bottom of the apparatus. After repeated failures, my grandfather, who was famously profane in his speech, finally exclaimed "Goddamit, *let* it slip!" When they did, he and his students found that they could much more effectively carry out the catalytic cracking at the bottom of the vessel in what is known as a fluid bed process. It is Doc Lewis's most famous engineering achievement.

Grandpa spent a lot of time thinking, talking, and writing about his conviction that engineering should become a true profession. As the decades have passed, however, it has become increasingly evident that the establishment of an autonomous engineering profession oriented toward ideals of broad social responsibility (the professional ideal of the little-e engineering) has not happened and is not going to happen. The social contradictions are too great, the epistemological core is not there, the institutional supports are not there, and the educational program does not begin to meet the stated goals. After repeated failures to achieve professionalization along these lines, it is time to admit that it is not going to happen, to let it slip, and to figure out another way of getting the job done.

It would pain my grandfather to admit this, but he often said that the one infallible sign of a disordered intellect is an inability to face the facts. He would also appreciate that professionalization is not an end in itself but a means of addressing the real problem, which his younger friend and colleague Elting Morison stated so succinctly: "how to organize a technological world we can live in."

In a hybrid world, engineering can thrive only as a hybrid. Today it is most dynamic at its peripheries, where it is most engaged with science and with the marketplace. Inevitably the profession formerly known as engineering will multiply into a much wider variety of grades and types and levels, because engagement with technology has far outgrown any one occupation. The future of engineering lies in accepting rather than resisting this multiplicity.

In terms of education, this means that the trend toward cramming more and more into the engineering curriculum runs in exactly the wrong direction. Few students will want to commit themselves to an educational track that is more and more nearly all-consuming. If a technologically grounded education is all or nothing, many students will end up with nothing. Instead, what we now call engineering education should be lowering the threshold for entry, mixing itself with the larger world rather than trying to keep expanding its own world. Today at MIT, and at many other institutions, students are trying to do this mixing on their own, but in too many cases they are trying to pour new educational wine into old institutional containers.

The most obsolete institutional container is that of the "engineering school." This can be a wholly separate institution, or it can be a relatively self-enclosed college within a university, with separate admissions processes and sometimes with a separate humanities and social science faculty. For example, many institutions (though thankfully not MIT) require that entering freshmen declare at the outset that they intend to enter the school of engineering rather than the broader-gauged school of arts and sciences. In the United States, these specialized institutional arrangements are a legacy of the nineteenth century, when new, technically oriented institutions defined themselves in opposition to the philosophy, pedagogy, and social assumptions of existing institutions of higher education. MIT's founder William Barton Rogers was, in the words of the Lewis Report, "motivated by a

deep-rooted conviction that the classic modes of higher education failed to meet fully the needs of the new era." Rogers
explicitly planned an institute that would be different. Professional and cultural studies would be mingled even in the undergraduate years, in an "integral plan" with an emphasis on
"learning by doing."[45] This would not be Harvard. This is MIT.

The raison d'être of an "engineering school" is to educate
students for engineering, defined as a distinctive profession
with its own well-defined identity. As this professional identity
dissipates in a process of expansive disintegration, engineering
schools will have to evolve or else find another raison d'être.
The segregation of engineering education served its purpose in
the nineteenth century, for it allowed an alternative model of
education to develop and flourish. Now segregation defeats
the purposes both of engineering education and of higher education, at once marginalizing engineering and depriving the
rest of higher education of its benefits. MIT still takes pride in
not being Harvard—but students at the two institutions today
are virtually indistinguishable in aptitudes (including verbal
and math SAT scores), in social background, in interests, and
in ambitions. Harvard, along with all the other Ivy League
schools and a host of other colleges and universities, is vigorously building up its technically oriented education, for which
there is enormous student demand.

The convergence of technological and liberal arts education
is a deep, long-term, and irreversible trend. The Lewis Report
extols the wisdom of MIT's "integral plan" as a new model
for higher education, summarizing its purpose with memorable brevity: "Education is preparation for life."[46] Students
need to be prepared for life in a world where technological,
scientific, and social and humanistic issues are all mixed
together, not just for engineers but for everyone. Such mixing

will not take place if students have to decide from the outset that they are attending an "engineering school" as opposed to a "non-engineering school." No matter how excellent the engineering school, and no matter how racially and ethnically diverse, if it attracts mainly faculty members and students who gravitate toward the technical problem-solving approach, then those students have an education that does not prepare them well for life experience. They need to be educated in an environment where they get used to justifying and explaining their approach to solving problems and where they get used to dealing with people who have other ways of defining and solving problems. Only a hybrid educational environment will prepare engineering students for dealing with technical "situations" in a hybrid world.

Engineering education must rejoin higher education in an adventurous mix that brings together information technology, the sciences, the social sciences, the humanities, and the arts. In a recent high-level review of engineering education, a coalition of leading engineering educators asked "What does the world now expect and demand of an engineer?" Their answer:

Students must be able to deal with ill-defined problems, to exercise judgment in the formulation of a task or in diagnosing faulty performance, able to tolerate ambiguity and work in the midst of uncertainty. For today's world they must know how to marshal evidence in constructing a rational proposal, describing the costs and benefits of one option compared to another and able to communicate that rationale to others. They must be able to put a variety of resources to use, especially the new computer and information processing tools, able to judge which tools are appropriate for each task they face. They should be able to work across disciplines, in teams with others who espouse different views, able to understand the non-technical forces that profoundly influence engineering decisions. They should be prepared for active, life-long learning, They too must have an ethical sense; recognize the legitimate role of code and regulation and know

the difference between cutting corners and rushed, but fully compe-
tent, effort.[47]

With the adjective 'engineering' omitted before 'decisions'
near the end, this statement would serve well as a statement of
general goals for higher education today. So why not think
about it as such?[48]

There are many obstacles to the convergence and remixing of
higher education. The pride and self-isolation of many engineering faculties is no more problematic than the pride and self-
isolation of many liberal arts faculties. Changes in attitude often
follow, not precede, changes in structures, however. As long as
engineering education remains structurally segregated, neither
faculty is motivated to work respectfully with the other. The
"idea of a university," according to the famous formulation of
John Henry Newman, is to be a "seat of universal learning"
where scholars "are brought, by familiar intercourse and for the
sake of intellectual peace, to adjust together the claims and relations of their respective subjects of investigation" so that "they
learn to respect, to consult, to aid each other." Newman continues: "Thus is created a pure and clear atmosphere of thought,
which the student also breathes, though in his own case he only
pursues a few sciences out of the multitude."[49]

MIT has grown in greatness as it has become more like a
university. There is every reason to believe that this will continue to be true in the future. As technology continues to spill
far beyond its nineteenth-century limits, institutions of all
kinds are disassembling and reassembling. At the aforementioned Zurich meeting about the future of engineering education, the historian of engineering Antoine Picon made an
offhand remark that "no institution can survive without inner
contradiction." Institutions associated with engineering will be
stronger, not weaker, for embracing those contradictions.

Advocates of a systems approach to engineering education see these trends, but respond to them with opposite instincts. Fearing that engineering might be marginalized, they want to make it even more comprehensive. Lockheed CEO Norman Augustine says that engineering should become socioengineering. More artistically, Thomas Hughes, the pioneering scholar of technological systems, has proposed that engineering become a *Gesamtingenieurkunst*, a Wagnerian grand opera where music, words, and stage design are all brought together in one grand trans-disciplinary production.[50] But this assumes that engineers are conducting the opera. The assumption is evident from the very act of naming them engineering or technological systems. Even if 'socio' is added, the terminology implies engineering's dominance in the world and implies that the heart of the system is technological. Instead of accepting the fact that the boundaries of engineering are dissolving, some systems advocates try to reestablish those boundaries, only further out than before. The universe swells larger, but it remains self-enclosed and dominated by engineers' habitual, often useful, but still limited concepts of problem solving. The standard of legitimacy still comes from engineering, whereas in the larger world—the common world—engineering has to make its way in the midst of different sets of knowledge, different forms of arguments, and different standards of legitimacy.

Hughes is an eloquent and loving critic of engineering who believes that "a major failing of engineering education and practice today is their failure to accept complexity, ambiguity, and contradictions" and that "the rigid rationality of engineering today verges on the tragic."[51] That rigid rationality will be overcome not by new ideas in engineering but by new people doing it. MIT has already made a good start in greatly expanding the ethnic, racial, and geographic diversity of its student body. It has also made a good start in greatly increasing the number of women in its student body. The supreme challenge for a place like MIT is to welcome more diverse intellectual interests among its students and its faculty members. Among the students this is already happening as a matter of fact: more and more of them are attracted to MIT as a place to get a technically oriented education that serves a basis for life plans other than traditionally defined careers in engineering or science. For many MIT faculty members, however, these students, however smart and hardworking, are a disappointment, because they do not share the focus of their teachers. Such students should be welcomed if engineering is to break out of rigid rationality and take its place in the larger world rather than defending a part of the world controlled by engineers. The potential tragedy here is not that of engineering, but that of the larger world, if, as a friend remarked to me, "all this technology seems to have nothing to do with being human."

"Engineers solve problems." What are the problems? The major ones are not of professionalization, not even of education, but of civilization in our new hybrid self-created habitat. In this habitat, neither the state nor the market has shown itself capable of managing global networks of technological and economic power. Innovations are being churned out at a furious pace, with few effective forms of governance for their synergies and consequences. New forms of political organization are taking shape, but it will be a long and arduous task to reconstruct democracy in ways that meet the needs of a world dominated by technology.

The identity crisis of engineering is much less serious than the identity crisis of democracy. They are related, of course. "Little-e engineering" wanted to claim an expert role in managing the things of the world ("facts"), while the politicians, for better or for worse, would manage the people problems ("opinions"). This division of labor never worked in practice and has now broken down completely. The scope of political decision making has to be expanded to the larger world of things. The French social thinker Bruno Latour puts it nicely: Politics has to be redefined as the "progressive composition of a common world." He calls this "Politics with a big P." Little p politics confines itself to non-natural phenomena—human interests and beliefs and other such residual, intractable, and insignificant matters. For the significant issues, now so much defined by science and technology, Politics is necessary. Engineering with a capital E has to get out of the way so that Politics can assume form and function, its battle cry being "No innovation without representation."[52]

Ken Alder, at the end of his study of Enlightenment engineers in France, writes: "Engineers were designed to serve."[53] Throughout history, the purpose of engineering has been defined by its patrons, not by engineers. The problems were posed by the state, whether from the civil side (aqueducts, bridges, tunnels, roads, buildings) or from the military side (weapons, supplies for armies, navigation). In modern times, the problems for engineering came from a mixture of the needs of the state and those of industrial capitalism, as filtered through the institutional needs of business enterprises.

The new master of engineering should be not the state, not the army, not corporations, not the market, but democracy. The strongest elements of little-e engineering culture are needed to support a political life that deals with the "progressive composition of a common world," which is just another way of saying a political life that asks "how to organize a technological world we can live in." To serve this master (and here is the irony, even the tragedy), engineering must both expand and disintegrate.

It must give up its quest for a well-defined professional sphere separate from politics and management. It must let go of its yearning for a segregated role of expertise and a segregated educational track. Engineering can save its soul only by losing it in the larger technological world.

As a support for democracy, engineering might be thought of less a separate profession than as a new kind of universal labor, as farming used to be when civilization was shaped primarily by its relations with non-human nature. Farming too has been an integrative enterprise, demanding a wide range of skills and knowledge. Though it has been practiced in diverse ways, its challenges have been similar enough to provide a civic connection based on shared life experience. In earlier days, the farmer was often regarded as the bedrock of American democracy. Can one imagine a Jeffersonian version of yeoman engineering? Can one imagine a vastly diffused engineering education that provides the civic basis for this kind of democracy?

In a conversation about how academic fields come and go at MIT, Bob Brown noted wryly that "back in 1970 we never worried about the death of anything." As we approached the new millennium, we kept noticing endings. It is appropriate to deliver an elegy for little-e engineering and for after-dinner conversations in a Newton sitting room half a century ago. Their time is past and will not come again. But little-e engineering has left a legacy that provides a solid foundation for the future: confidence in the ability to solve problems; energy and hard work; the habit of teamwork; the understanding that dedication to a job can be as important as brilliance brought to it; the delicate balance between investment in and distance from capitalism; the belief in community and progress; the assumption of responsibility for the material world. All these provide a technology-based alternative to the ideology of technological change. The world of little-e engineering may have been narrow and cramped and lacking in epistemological clarity. It compensated for these deficiencies, however, with a well-defined code of behavior and sense of public service.

And so we are not confronting the death of engineering any more than we confront the death of nature, history, or science. All these categories are being transformed simultaneously and interactively. For individual engineers, however, it will be painful to make the transition from acting in a familiar world over which they have considerable control to acting in the larger and more disorderly world. It will be especially painful when technology itself—innovations, knit into systems— intrudes into the familiar world of engineering. This is what happened when MIT committed itself to Reengineering.

# 3

# Technology and Business

## The Waves of Change Wash on Shore

When I first became involved in Institute governance in the early 1990s, at nearly every meeting, it seemed, the president or the provost displayed the same ominous graph: a line chart of MIT expenditures and revenues over the past twenty years. From left to right (past to present), there was an ever-growing gap between the top line (spending) and the bottom line (income). If the operating gap was projected into the future, MIT was headed for the financial rocks.

At the time, nearly all institutions of higher education were facing a similar deficit. In a service-intensive industry, the actual cost of educating students was rising faster than income from tuition and fees. Government support for student financial aid was declining, especially on the federal level. MIT was especially vulnerable because Cold War research had been especially good to it. In the early 1990s, direct federal support for research was at best flat, with substantial cuts threatened, and indirect cost recovery was being cut dramatically. (Just one change in rules about support for graduate student funding cost MIT $15 million a year.) "We are so highly leveraged, it's scary," another dean remarked to me in those anxious days.

MIT had to get spending and income more in balance. The most obvious remedies had already been tried. When the first round of the budget gap occurred in the 1980s, MIT cut its non-faculty work force; within a few years, however, the work force was back to its former size. In the early 1990s, the MIT administration tried three consecutive years of 2 percent budget cuts. By the third year it was clear that the slopes of the income line and the spending line were not getting any closer. MIT urgently needed to find a better way to steer clear of the financial rocks.

President Vest responded with a three-prong strategy: advocating for continued government investment in research, developing new sources of revenue, and improving management of existing resources. To achieve the latter, Vest announced in November 1993 that MIT was committing itself to the Reengineering Project. The goals of the project, he explained, were to have MIT operate more effectively while reducing its operating gap by $40 million ($25 million net of indirect cost recovery) and its work force. Using the budget numbers, people commonly figured that might mean cutting 400 jobs of the roughly 10,000 staff positions.[1]

"Reengineering" was one of the most popular management techniques of the day, and it seemed to have particular affinities with MIT. In its official announcements, the administration described reengineering as "the fundamental rethinking and radical redesign of support processes to bring about dramatic improvements in performance."[2] In other words, reengineering turned the deficit into a design problem. This seemed, in President Vest's words, "MIT-like." Vest keenly wanted MIT to develop, in his words, "its own style of Reengineering." Reengineering should fit well culturally with MIT, he explained, since MIT people like to take things apart to see how they might be made to work better.[3]

Vest was right about the people, but events were to prove him wrong about the cultural fit. At MIT in the 1990s the most significant "culture war," by far, was that between engineering and Reengineering as different incarnations of "technology." For many, MIT's Reengineering Project was defined as a struggle between technological innovation and cultural resistance in the workplace. This is a familiar story of the late 1990s, and it continues to be repeated today in countless settings. The surprise, and the interesting thing about this story, is that the struggle was acted out at MIT—of all places on earth, one would think, among the least likely to resist technology. The fact that it did happen at MIT is a clue that the real story is not one of technological resistance.

Many MIT employees were torn between the existing world of work at MIT and the "new world of work" launched by the Reengineering Project. Most of them could not avoid this conflict, because of the administration's decision to focus redesign on "support processes." This decision deliberately shielded core faculty processes such as research and teaching from redesign. Nevertheless, many MIT faculty members denounced Reengineering. At one dinner attended by MIT faculty members, an engineer suggested that Reengineering be called "resciencing" because it gave a bad name to engineers. In other conversations scientists were just as vociferous in denouncing Reengineering as anything but scientific and in blaming Reengineering on engineers. Faculty members were largely united, however, in objecting to the high cost of Reengineering and to its managerial rhetoric ("It makes me feel like I've been abducted by aliens," one told me). The resentment was compounded by the predictable tendency of decentralized administrators to resist greater control by central administration.

In retrospect, much of the opposition could have been avoided if the "change process" (Reengineering lingo) had been presented much more narrowly from the start as the implementation of a new accounting system. Over time, Reengineering became practically synonymous with the introduction of new financial software, SAP R/3, which replaced MIT's general ledger, accounts payable and receivable, and procurement systems. As an institution with 10,000 students and as many employees, MIT had approximately 100,000 separate financial accounts, which had proliferated to meet the needs of various departments. The systems managing them were poorly coordinated, inefficient, and aging.

Even the worst enemies of Reengineering would admit that MIT needed to replace and integrate its accounting systems. Over the six fiscal years of the Reengineering Project (1994–1999), one-time costs totaled $65.2 million, and of this, $41.8 million went to upgrading financial systems through SAP.[4] It was a huge investment, but not out of line with those made by other universities. During the 1990s, research universities similar in size to MIT collectively spent about $500 million on accounting software. It had to be done; however, it should not have been expected to reduce costs, and did not have to be connected with the overblown rhetoric of technological change.

Reengineering was a management fad, a transient episode, a part of the wave of consulting panaceas of the early 1990s. That wave began to form in the 1980s, when businesses began to use software systems in large-scale and strategic ways to restructure themselves for greater integration and greater flexibility. For the most part, those software systems had not been invented to meet the needs of business. Information technology originated in a much more interesting and unpredictable mix of circumstances, including military needs, 1960s culture, regional dynamics in the San Francisco Bay area, and the internal momentum of technological discovery and development. But when all these circumstances converged in the 1980s to make robust software systems available, those systems began to be used and shaped by corporate restructuring projects. If anything, MIT and many other universities were late getting into the reorganization game.[5]

Many of the management fads quickly dissipated once the "new economy" got up and running in the later 1990s. However, the drive to restructure work and organizations with the aid of new software systems continues with as much energy as ever at MIT and elsewhere. The headline above a valedictory report published in the MIT Faculty Newsletter at the end of 1999 announced: "Reengineering Is Over but Change Is Not."[6] The wave of Reengineering has been overtaken by the swelling tide of "change management," with no end in sight.

As dean, I often struggled to manage an office where the cultural tensions raised by Reengineering constantly undermined any functional advantages it offered. As a historian of technology, however, I was fascinated by Reengineering. From it I learned two major lessons about technology that I had not learned as a historian.

The first lesson is that distinctions between matter and mind, between technology and society, and between engineering and management are gone. I had known this intellectually, but I now came to know it in a much more intense and visceral way. It is as hard for historians of technology as for engineers to let go of the assumption that technology is material—that it means bridges, steam engines, airplanes, and all the other machines and structures that fill the human-built world. But in Reengineering meetings we would talk about "feeds" and "bridges" or "drilling down" and "bolting on" this or that, and we were always talking about "code"—about social connections, not material ones. The problem we were trying to solve was not anything posed by nature: it was a financial crisis. If Reengineering was a design exercise, it was about redesigning work, not matter. Was this engineering, or management? Who could tell the difference? Members of the MIT faculty might disdain Reengineering for giving a bad name to engineering; however, for the MIT staff Reengineering defined what "technology" is in this day and age.

This is the second lesson I learned from Reengineering: that many engineers do not understand this interpenetration of technological and organizational change, and consequently do not understand their own role in history. The ideology of little-e engineering was still strong enough that many engineers reacted in dismay against the ideology of technological change that came with Reengineering. I had conflicting feelings about their reaction. Because I respect the moral culture of engineering, I was heartened by their resistance to the latest version of "management bullshit." On the other hand, this resistance made me uncomfortable, because on some level it was a refusal to face the facts. One significant fact it evaded was that non-faculty employees had to take Reengineering seriously, even if faculty did not.

Even more deeply, many on the engineering faculty were refusing to face their own historical role in providing the tools for such "change processes." For decades, MIT researchers have been working on these tools. Their inventiveness had been critical in the development of technical systems to manage the work of large organizations—first military systems (SAGE,

ARPANET) and later civilian organizations (airline operations, government data processing, Great Society projects).[7] It was only a matter of time before the waves of creativity and enterprise going forth from the Institute came back to its shores. What came back seemed unfamiliar, because innovations that had left the Institute as discrete machines and devices returned as complex systems, shaped by entirely different contexts and purposes. Although often developed and marketed by MIT graduates, these systems of control and communication seemed unfamiliar and threatening even to those who contributed so much to creating them in the first place.

In the language of Marxism, this is alienation—the separation of workers from the products of their labor, or, in more contemporary language, of innovators from innovation. According to Marx, alienation is inevitable in a capitalist economy. It is more than an indictment of capitalism; it is a human tragedy. Once rationality is reified (another key word in Marxism) in the form of things, it can imprison and eventually subdue the creative vitality that generated it. This is part of the problem of "how to organize a technological world we can live in." To begin, we have to recognize that this is a world we have created. In its disdain for Reengineering, "little-e engineering" did not sufficiently acknowledge the reality of a reflexive world in which waves of change come back to the settled shore in unpredictable and sometimes unwelcome ways.

## O Brave New World

The Reengineering Project got underway in March 1994. An eight-member Core Team reviewed ("mapped," in Reengineering language) major support processes of the Institute and eventually recommended that eight be redesigned: facilities operations, mail service, supplier consolidation, management reporting, information technology, appointments, human resources, and student services. When I accepted the position of dean, in the spring of 1995, I understood that a major part of my job would be helping to lead the redesign of student services. In order to do this, I was made a member of the Reengineering Steering Committee, which met weekly or biweekly to exercise general oversight over all the major reengineering teams.

From 1996 to 1998, along with many of the approximately 450 employees who reported to me as dean, I found my working life oriented around Reengineering. It was a heady time. For staff who had long been laboring at MIT, and who had long been thinking about how things could and should be done better, Reengineering was an opportunity to imagine, to think big, to think boldly. Most employees felt that an examination of the "fundamental processes" of student services was long overdue. They were aware that students had to shuttle back and forth across the campus to take care of basic transactions. They felt they were doing the best they could with the physical space and information systems they had. They were excited to contemplate better space and systems.

The Student Services Redesign Team was asked to review student services processes and to consider fundamental changes. Team members had to give up their regular jobs for some months so they could focus their time and energy on planning for the future rather than maintaining the present. This put pressure on those who kept things running—an inevitable outcome of a culture that values innovation more than maintenance. In some cases, jobs were "backfilled"; in other cases, co-workers pitched in with overtime to cover the work of team members.

In the spring of 1996, the Student Services Redesign Team emerged from several months of "visioning" with a list of ten major areas it recommended for further work. In almost all cases, its proposals presumed a fundamental change in the culture of student services at MIT, from one that was hands-off and laissez-faire to one that was more supportive and communitarian. The hands-off culture was deep in MIT's history: it was part of what had defined MIT as not being Harvard. Over the years, for many reasons, including the lower-middle-class origins of many MIT students, student life evolved as a refuge from the academic rigors of the Institute, rather than as an integral part of the educational experience.

The redesign team felt that the division between student life and learning should be overcome so that the overall educational experience would be stronger. Students, they felt, needed more opportunities for informal interaction with faculty members and other adults, and more support for their campus and residential activities. This fresh approach would include, for example, a more integrated set of counseling services, a team-based advising system, and a freshman orientation focused less on residential choices and more on academic choices.

To do any of these things required more than new "processes": it required a new or at least modified cultural understanding of student services at MIT, which in turn would require long and substantial discussions among students, faculty members, and non-faculty employees. But Reengineering was business oriented, and it was impatient. Reengineering consultants wanted quick results, commonly defined as "low-hanging fruit"—process changes that would have high visibility and could be accomplished quickly. Some Reengineering leaders openly proclaimed the need to "break the institutional culture" of MIT so a new one could flourish. Their interest was in the culture of staff work at MIT, not in faculty or student culture. Their language was often utopian: Reengineering was building a "new world of work" or simply a "new world." The fundamental change they wanted was not in educational experience but in staff self-identity.

At a university, however, staff self-identity is defined through a delicate balance of human relationships among staff, faculty members, and students. In the prevailing MIT culture, the most valued staff were not necessarily the most brilliant or even the hardest-working but those who knew their units well, who performed consistently over the years, and who showed a high degree of loyalty to the people they worked for and to the educational mission of the Institute. The faculty members were supposed to be dynamic, mobile, self-driving leaders in their fields; the role of the staff was to provide stability and support for them. You could describe the equilibrium as one balancing male and female roles. You could also describe it in geographic terms, as an equilibrium between faculty members plugged into global networks and staff providing situated, thick, local knowledge.

Reengineering upset this equilibrium. It promoted a staff culture much more like that of the faculty: dynamic, self-promoting, engaged in lifelong learning, mobile, aggressive. Now staff too were encouraged to become self-creating, self-aware "change agents," relying on a suite of skills and "competencies" for long-term security, instead of looking for security in the institution employing them. It was a theatrical view of the working world: staff were encouraged to think of themselves as "performers," and a job was defined as "the combination of roles a person performs." One staff member told me, with considerable enthusiasm: "We'll all be in theatrical production; there is an artistic aspect to all this." Personal identity, not institutional loyalty, was the new basis for career development, which, he noted, looked like a "series of gigs." The utopian vision of self-reliant, self-empowered, endlessly flexible workers was, in sum, a staff version of the entrepreneurial dream so attractive to engineering students today.

Much of the work of Reengineering went into promoting the "new world of work" at MIT. Over platters of bagels, fruit, and sandwiches, team participants were guided by consultants through a wide array of cultural practices that the MIT staff found (depending on the person and the practice) amusing or offensive, useful or degrading. Team-building exercises involving rubber balls, "stickies" massed on poster paper, facilitators positioned next to flip charts. Team members would return from day-long retreats devoted to "visioning" exhausted but exhilarated, with piles of flip-chart leaves, studded by stickies, on which were sketched process maps and preliminary plans for "process redesigns."

Since I taught writing at MIT, I was particularly intrigued by the rhetoric of Reengineering. Like most cultural revolutions, it insisted on a new vocabulary. It is almost too easy to make fun of this rhetoric, which had a serious purpose: to help usher in the new world by providing a new language to explain to ourselves our new social environment. One staff member marveled: "Five years ago I never heard of facilitation, and now it's all we do!" Students became "customers" and faculty members "clients." Staff lost their specialized titles and were given generic ones: "process owner," "team leader," or, of course, "change agent." In job interviews and performance reviews we focused on "competencies," a term that covered both personality traits and specific skills.

Just as significant as the shift in vocabulary was the shift in genre (so to speak) from paragraphs to lists. At most of the MIT meetings I have attended over the years, verbal communication has taken the form of discussions in which various individuals have contributed a minute or two of commentary to make a point. Written communication has taken the form of reports that are essentially persuasive essays, with introductions and conclusions and supporting paragraphs. At the Reengineering meetings, however, facilitators urged participants to make lists; discussions took the form of prioritizing lists; reports too took the form of lists. The genre of bulleted items came to rule, along with presentations in PowerPoint (a program designed to support communication by lists).

The pervasive problem was extracting the point from the PowerPoint, the argument from the lists. In one session of the Reengineering Steering Committee, after a particularly florid ten-minute presentation of multi-color flow charts and pie charts with multiple 3-D slices, there was a long silence after the projector was switched off. We were all astonished by the display, but after the silence one team member ventured to ask "What did that all mean?" We took another ten minutes to figure out that we had learned how to order paper clips through the new supply-procurement system.

Even when you are part of a well-established bureaucracy— maybe especially when you are part of it—you can dream of smashing through it to something simpler and purer, to achieving an empowered state of individual freedom. More than once, in team meetings where we were surrounded by flip charts bulleting the vision of the new world, I recalled Wordsworth's lines written in the early years of the French Revolution:

'Twas bliss in that dawn to be alive
And to be young was very heaven.

But Reengineering, like all utopias, was built on illusion. In an age of complicated systems, both organizational and technological, it is an illusion to think that self-activating individuals can sail alone through the seas of change. Technology needs culture. If the culture of the Institute was going to be "broken," another would have to be built in its place. As early as the spring of 1996, one staff member who had been a strong participant in the Redesign Team remarked to me: "Reengineering is becoming just another bureaucracy." Teams proliferated, budgets grew, procedures multiplied. The steering committee failed to steer, revealing a structural dilemma of leadership in the information age: the people with the broadest and deepest experience of organizational culture are not necessarily knowledgeable about information technology, while the consultants most savvy about "the technology" are not knowledgeable about the organization. In fact, the interest of the consultants lies in dismissing this knowledge as irrelevant: they will move on, and their interest lies in emphasizing supposedly contextless software.

This is a fundamental reason for the failure of President Vest's hope that MIT would develop its own style of Reengineering. The people most attuned to MIT's "style" were not necessarily effective at managing people with SAP expertise, while SAP experts had little understanding of how MIT works. Staff (faculty and non-faculty) were confused about who was in charge. Instead of a brave new world of individual empowerment, they felt yanked back and forth between the old world of work at MIT and a promised new one. Worse yet, if they tried to defend MIT against what they considered destructive changes, they might find themselves accused of disloyalty. Reengineering presented some people at MIT with a deep moral dilemma: they had to find the courage to deliver warning messages knowing that the reaction might be to blame the messenger.

Conflicts had been set up from the start, in the mixed signals that were sent out with the announcement of the Reengineering Project. Employees who were asked to join Reengineering teams were invited to join in a form of participatory management. They were being asked to reflect on their work, an opportunity that led to a good deal of enthusiasm and even moments of euphoria. But while this increased self-awareness was liberating, it was also intimidating, because it took place in the context of a budget gap. While Reengineering told people to challenge how things were done at the Institute, the Institute—which had not gone away—reminded them that their jobs were not secure. In the kickoff announcement of November 1993, two goals had been highlighted: reduction of the operating gap and reduction of the work force.

Reengineering was essentially an exercise in "self-Taylorizing,"[8] and as a result it caused painfully split identities and split loyalties. Employees who were also members of Reengineering teams were never sure if their participation in Reengineering was just a way to provide management with cover when deciding who would lose his or her job. Throughout the Reengineering Project, they were subjected to confusing switches between tough get-on-board talk and sentimental feeling-your-pain-of-change talk. The therapeutic language of employee self-empowerment was received with cynicism, because staff understood their jobs might be at stake. In the end, large-scale job cuts never came, but no one could have known that in 1996 and 1997.

There is no easy way to shift from a world of work defined by an institution to a world of work defined by personal identity. MIT people—all of them—were caught in the middle, living with confusion and conflict. The Institute had made it clear that non-faculty employees were expected to cooperate with

Reengineering, and many of them wanted to turn it to good advantage. On the other hand, many faculty members were critical of Reengineering, and staff needed and wanted to keep them happy too. This led to an endless series of compromises and negotiations as non-faculty employees tried to straddle the fault line. To do so, they invented an array of pidgins and creoles. They were constantly trying to translate Reengineering-speak into language that would be comprehensible and meaningful to the rest of MIT. In one email exchange on which I was copied, staff tried out the idea of consciously varying their language so faculty members would be addressed in "scientific terms" while administrators would be addressed in "plain speak" or "jargon," depending on the personality of the administrator. "Our language has become a barrier to success," lamented one emailer who referred to the language of Reengineering as "corporate yuk." The choice of language was telling, but other choices were even more wrenching. Would you join a Reengineering team? Would you make a presentation about Reengineering to a faculty group? Would you openly criticize Reengineering, or would you keep your criticism to yourself?

Over the years, MIT has had an enormous advantage in hiring and keeping staff because its employees believe that their work is significant, and because they get to work with compatible and interesting people. Such motivations are a source of efficiency. Trust and loyalty reduce the need for bagels, retreats, and facilitators. Trust and loyalty allow people to make demands on one another, to have high expectations of one another. Trust and loyalty allow people to come to terms with technological change when they know that the larger goals and the social setting are relatively stable. This was especially noticeable in academic departments, where the core culture of MIT has developed over generations of interaction among non-faculty employees, faculty members, and students. In this local setting, the efficiency of the organization depends on the willingness of people to undertake multiple roles based not so much on work processes as on human relationships. Reengineering had to perform an immense amount of cultural work to replace the hidden efficiencies of the old institutional culture.

As a part of the old culture, I was not eager to see it broken. Nor did many outside of MIT. While Reengineering was trying to "break the culture" and create a "new world," a stream of visitors kept flowing to MIT to admire the existing institution and to learn how they might emulate it. Nearly every week, senior administrators got requests to meet with such visitors. MIT entered into several high-profile agreements with other institutions that wanted to imitate its success, and it turned down many more such invitations.[9] The ideology of change management disparages institutions as obstacles to the flow of innovation. Yet, of all the inventions that MIT has produced over the years, none has been as influential as MIT itself.

The emphasis on identity-based work implies a reduced attachment to institutions. For all their therapeutic language, institutions are not asking workers to take more personal responsibility for their careers in order to liberate the oppressed. Instead, by demanding this they hope to attain more flexibility in hiring and retaining workers. But that increased flexibility comes at a price, both for workers and for institutions. Institutions impose constraints, but they are also empowering. The idea of working at a place committed to a significant goals—making scientific discoveries, solving problems, and educating the next generation—is usually more empowering than the idea of starting an endless "change journey" with no evident destination. I have yet to encounter someone who truly gets excited about contributing to "change."

Humans are deeply conservative creatures, not necessarily in a political sense, but in a cultural sense. The social scientist Peter Marris, in a study of experiences of loss and change, including bereavement, notes that "the impulse to defend the predictability of life is a fundamental and universal principle of human psychology." As "adaptive beings," our tendency is "to assimilate reality to [our] existing structure, and so to avoid or reorganize parts of the environment which cannot be assimilated."[10] We require a "structure of meaning," in the sense of "organized structures of understanding and emotional attachments, by which grown people interpret and assimilate their environment."[11] In this sense, the conservative impulse is "as necessary for survival as adaptability: and indeed adaptability itself depends upon it."[12] When predictable patterns of life are disrupted, we can come to terms with disruption if we are able to take hold of a thread of continuity, to interpret what has happened in a way that reconnects us with what has been lost, and to "repair the thread, tying past, present and future together again with rewoven strands of meaning."[13]

Reengineering glorified "the change process," but it systematically cut the threads of meaning that might have helped people at the Institute adapt to change. The most powerful lesson of Reengineering is that a "new world" may not be a better one. In resisting Reengineering, people at MIT called upon the past as a better world, as a resource for resisting change for the worse. MIT has always identified its mission with the future. Defending the institution in the name of the past was an unfamiliar exercise.

Many meetings of the Reengineering Steering Committee were held in conference rooms in MIT's Main Group, the monumental set of Beaux-Arts-style buildings that branch out below the Great Dome. During PowerPoint displays, I would sometimes muse on the resemblance between the Main Group and the palace of Versailles. I looked out on Killian Court, a green rectangle stretching in front of the Great Dome, half expecting to see a guillotine set up there to dispatch old-world administrators like myself. I began to think less about Wordsworth and more about Edmund Burke, who denounced the French Revolutionaries for their utopian illusions that destroyed the collective wisdom of institutions that had been built up slowly over many years.

### Technological Drift

MIT's official definition of Reengineering—"the fundamental rethinking and radical redesign of support processes in order to bring about dramatic improvements in performance"—said nothing about focusing the redesign on new technologies. In the spring of 1996, when the Student Services Redesign Team recommended ten processes for further work, follow-on teams were set up to explore redesigns for five of them. Many of the recommendations proposed increasing face-to-face contact among faculty members, staff, and students, in order to try to reduce the barriers between student life and learning.

From the long and rich list of possible improvements, Student Services Reengineering inexorably drifted toward the ones dominated by technology. Even before the Student Services Redesign Team filed its report, Reengineering leaders had made an independent decision to invest heavily in consolidating the work of the three most transaction-intensive offices: those of the Bursar, Student Financial Aid, and the Registrar. This decision was made outside the team process because it seemed self-evident that delivery of these transactions could be improved dramatically: they were, as the saying went, "low-hanging fruit." Personnel from the three offices were selected to serve on the FAST (Financial and Academic Services) Team, which got fast-track funding to move forward with implementation while other areas of student services were in the planning phase.

Before long, the FAST Team and its successors dominated Student Services Reengineering. The three existing offices were consolidated to two. A new technical support unit was added in the form of an office for Student Services Information Technology. These organizational moves were accompanied by physical ones. Most members of the Financial and Academic Services staff moved to a single building, which was extensively renovated. As part of the renovations, a comprehensive Student Services Center was given a prominent place on MIT's "main street," the so-called Infinite Corridor. Instead of having to walk around campus to take care of financial and registrarial chores, students could go to the Center for "one-stop shopping," where they were assisted by cross-trained staff. Even more important, they could do much more "online shopping," accessing their own financial and registrarial information without any staff assistance. They could check their financial aid status on line instead of asking a financial services staff member. Instead of filling out paper forms, they could make electronic applications.

The Student Services Center was a Reengineering showpiece. It became a regular stop on tours for potential applicants and visitors. The senior administration regularly and justifiably praised the work of the FAST Team as an example of the best of Reengineering. The staff deserved this praise, and as a manager I was happy to hear it. As a historian, however, I was troubled by how quickly and definitively a wide range of desirable changes had been reduced to *technological* changes.

I thought about this as an example of what the historian Thomas Hughes calls "technological momentum" in history. Once a large technological system is in place, Hughes says, its scale can acquire momentum, a quality "analogous . . . to inertia of motion." Such systems "have a mass of technical and organizational components; they possess direction, or goals; and they display a rate of growth suggesting velocity." They are not in fact autonomous, but they can appear to be.[14] Like many other historians of technology, Hughes leaves considerable room for human intervention in the design stages of technology, especially in what he calls the invention and innovation phases of technological systems. Once the designs are fixed and built, however, it is difficult to alter their trajectory.

What I saw in Reengineering were the very earliest stages of technological momentum, way back in the design or even predesign stages. In the Reengineering Project, technology was identified with change in the earliest conceptual phases: technological changes were the low-hanging fruit, and so technology came to drive the change process by defining what kind of systems were to be built in the first place. This initial drift toward technological solutions in design set the stage for the more evident technological momentum in implementation.

In Reengineering, technological drift was motivated by utopian illusions. Recall that the basic idea of Reengineering was to "redesign" work. If it had been said that redesign would make work more efficient and effective, we would have been working with the language of improvement, of attainable goals. As we have seen, however, Reengineering had a strong utopian element: it promised not just improvement, but innovation, a new world, a new culture, "the *fundamental* rethinking and *radical* redesign of support processes to attain *dramatic* improvement in performance."

How do you get *dramatic* improvements in student services? It is notoriously hard to increase productivity in services, where the quality of human interaction is crucial and where technologies offer limited advantages.[15] Whether in the classroom or in residential and campus activities, education requires ongoing, daily, intensive human interactions between students and older adults: coaches, deans, professors, staff. When a student comes in to talk with a staff member about loans or financial aid or a part-time job, they often end up discussing a wider range of issues—academic problems, family struggles, career plans—which are often mixed in with financial issues. A university thrives on hybrid conversations in which informal counseling and benign surveillance are mixed in with the delivery of information. By their very nature, these conversations are inefficient in the sense that the purported topic may not indicate its real substance.

If transactions become too efficient and autonomous, the university community loses opportunities for conversations. When the registrar's office began to put class registration on line, some faculty members protested that they wanted to make sure this was done in a way that maintained the ritual of Registration Day—the day before classes start for the semester, when students meet with faculty advisors to get their registration forms signed. Like most rituals, this one can easily be romanticized. Some faculty members have too many advisees to spend much time chatting with each of them, and sometimes the personal chemistry does not work. But more often than not, the ritual of registration provides an opportunity for face-to-face communication. Many faculty members resisted any new form of registration that removed this opportunity for conversation.

When student services have a mixed pedigree—part transactional, part conversational—the transaction is usually the cheap and easy part. For example, in arranging the annual student housing lottery, we used ever more sophisticated algorithms, so that more and more students got their first and second choices of residence hall. The lottery system kept improving, but this had no evident effect on the really significant issue of students' happiness with their residential life. Dean's Office staff and faculty housemasters continue to struggle with difficult and time-consuming housing problems that inevitably arise no matter how successfully the lottery works: roommates who do not get along, noise on the corridors, family emergencies that require unexpected moves. In many of these situations, students need the help of older adults, and we have not found a way to speed up dispute resolution. Indeed, the number of students in distress seems to be increasing, for reasons that are not clear and that for the most part are beyond the control of the university. In fact, the best and maybe the only way to achieve *dramatic* efficiencies in student services is through *dramatic* change in admissions policy, so that only the least trouble-prone applicants are admitted—but this is not an acceptable option, for many good reasons.

Reengineering was committed to *dramatic* improvement. This could be accomplished only by transforming "processes" into transactions, which could be defined and addressed in technological terms. In the world of change management, where major improvements are promised, it is as natural for water to run downhill as it is for change to flow into technological channels. As MIT's senior vice president once remarked, "We're in a world of rising transactions." The tide keeps rising as problems are framed as transactions to be handled with software. Not just handled, but eliminated: the ultimate lure of change management, the "technological mirage" down the road of reengineering,[16] is the promise of "making the work go away" (a phrase used often in Reengineering circles). With the right software, we were told, some work would no longer have to be done, at least not by a paid staff employee (the customer might do it for himself or herself). By promising that "going electronic" would make work "go away," Reengineering went beyond utopia to magic, beyond a new world to a magical one where some work simply vanished. As a result, the reengineering of Student Services was characterized by technological drift. Non-transactional or only slightly transactional services, such as freshman orientation, were not reformed at all, in contrast to the considerable resources put into the FAST Team. Services of mixed nature were supported in their transactional dimension but not in other dimensions.

It was in genuinely transaction-intensive areas that Reengineering had its greatest and most positive effect. Even in these areas, however, the work did not magically disappear. If work decreased in one place, it popped up in another. Reengineering provided numerous examples of the reflexive nature of the technological world, where a change causes reverberations that act back on the change itself. We created information systems intended to save labor in an existing context, only to have that context evolve as people responded to new information systems. For example, when students and their families could access their student aid information from the Web, they did not stop contacting financial services staff. They started emailing as well as calling, they asked more sophisticated questions, and they bargained more aggressively.

Similarly, the hope of making work go away by making printed documents go away proved a mirage. Putting information on the World Wide Web rather than printing it redefined work but did not eliminate it. Although it takes time to get publications ready for the printer, once they have been published you can forget about them for a while. In the words of one leader of the new Student Financial Services Office: "The paper used to shield you." Web-based information requires continual updating, monitoring, and revision. The transition from paper to email only intensified the communication process.

In addition, "going electronic" or "going digital" (two other phrases often used in Reengineering) created at least as many jobs as were eliminated. A new unit, Student Services Information Technology, was needed to develop, implement, and maintain new software. Overall, we needed not fewer staff, but different kinds of staff—notably programmers, database administrators, and webmasters. Furthermore, adding software did not readily translate into reducing staff. Even when "going electronic" eliminated work, it usually eliminated only a portion of a job. Even when added up, these portions of jobs rarely made it possible to eliminate a position. It was difficult to "harvest" the human savings.

In trying to manage these changes, I often found myself thinking of Ruth Schwartz Cowan's book *More Work for Mother*, a classic study of failed technological magic—the mirage in this case being the illusion of eliminating housework.[17] Cowan shows how the promise was subverted by rising expectations and by the inherent limits to the kinds of work that technology can make more efficient. As Cowan shows, redesigning work becomes redefining work. The housewife did not do the same list of tasks more efficiently; the list changed, and the identity of the housewife changed. Work rarely disappears; it constantly mutates.

The larger MIT experience with Reengineering confirms the tendency toward technological drift. To be sure, there were some dramatic examples where work was eliminated or made more efficient through non-technological means. For example, MIT eliminated its Graphic Arts unit, preferring to outsource most of that work. Repair and Maintenance was changed from a centralized to a zone-based system. Mail Services eliminated delivery to individual offices in the parts of campus that had it, instead requiring "customers" to pick up their own mail from Distributed Mail Centers. But once SAP implementation was fully underway, technological drift became a vortex, and Reengineering became synonymous with the new software system. For MIT at large, as for student services, there was no technological magic making work disappear. "Even though Reengineering is over," the semi-official summary of Reengineering in the Faculty Newsletter reported, "MIT is still very much in transition. Tools such as SAP have not been easy to use, and the learning curve has been steep. In fact, some administrators would probably tell you they have more work now than they did before Reengineering began."[18]

It is just as true in academic departments as in administration that hybrid work—part transaction, part conversation—is the rule rather than the exception, and that individuals are often valued most for their ability to perform multiple roles. But even in the academic departments, technological drift tended to highlight only the transactional part of the work. At one meeting about administrative information technology, held after the official end of the Reengineering Project, an administrator complained: "The balance of the work is shifting to the computer. Instead of paddling around talking to people and putting out fires, now the administrative officers have their nose pinned to a screen. They need to keep sparring faculty members apart and pay attention to the postdocs. They don't know any longer what is going on. They confuse having information with doing their job."

At the aforementioned meeting, we discussed how individuals' jobs had been redefined by SAP R/3. "Do people know what to do when they are asked to review the information?" one participant asked. The reply was: "They should think about it, use their judgment, but sometimes they just look at it. They don't know what to do with it." We concluded that we needed to do a better job of educating administrative officers in using judgment to interpret the implications of the reports, and to pay attention to their non-transactional roles.

Once work is reconfigured in technological terms, however, it is profoundly shaped by the logic of the supporting technological system. The rules that govern the technology start to govern everything else. Technological drift becomes technological momentum, which begins to feel very much like technological determinism: a point in history where machines have come to control people, not the reverse.

### Technological Determinism

For historians of technology, technological determinism is the unthinkable thought. First, it is self-evidently untrue: human beings construct machines, not the reverse. Worse, the thought is dangerous: if we start believing that machines run us, we will give up what control we have. According one historian, belief in technological determinism "inhibits the development of democratic controls of technology because it suggests that all interventions are futile," and "if we do not foster constructivist views of sociotechnical development, stressing the possibilities and the constraints of change and choice in technology, a large part of the public is bound to turn their backs on the possibility of participatory decisionmaking, with the result that technology will really slip out of control."[19]

When I became involved in Reengineering, however, I heard my MIT colleagues express this unthinkable thought every day.[20] A few examples:

We have no choice but to go with the upgrades.
Email is killing us, and there isn't a thing we can do about it.
We have to adjust MIT culture to fit the software.

My ungrammatical but expressive favorite, from someone deeply familiar with financial aid software, is this:

The software puts a whole new structure of thought in front of you that people have to adjust to.

As a historian, I winced to hear my MIT staff colleagues express undemocratic capitulation to technological determinism. As a manager, however, I had to agree. I could not think of a way to avoid the upgrades, however unnecessary, or to avoid adjusting how we did things at MIT to fit the dictates of the software. I certainly could not find a way to avoid the flood of email. Most MIT staff employees and not a few faculty members—generally smart and thoughtful people—are convinced that technology does play a determining role in their lives, and they certainly do not think it is under their control. Are they suffering from collective hysteria, from a mass attack of false consciousness?

Thomas Hughes, in his analysis of technological momentum, distinguishes between technological determinism as theory and technological momentum as experience. It is easy to refute the logic of technological determinism, but the everyday experience of having to conform to "the technology," "the software," or "the computer" cannot be refuted by logic. From everything you can see, the machine tells you what you can or cannot do. In the cybercapitalist version of magical realism, the narrative (events of daily life) floats free from logic (an explicable sequence of cause and effect). Unlike computers, human beings are only partly logical. Much of our thinking is shaped by images and stories. This is not false consciousness, but human consciousness. Even if the theory of technological determinism can be refuted, the experience of technological inevitability is convincing.

As I have already mentioned, the first lesson Reengineering taught me as a historian is that the distinction between technology and society is truly gone. Once I had absorbed this lesson, I understood that technological determinism is a non-issue, because there are no technological forces separate from social ones. Of course technology is socially constructed! How could it be otherwise, "since technology *is* society, and society cannot be understood or represented without its technological tools"?[21] The real question to ask, then, is not "Is technological determinism true?" but "What are the historical forces shaping the construction of the technological world?"

For all the headaches it caused, Reengineering allowed me to follow the process of human beings trying to "organize a technological world we can live in." The transition from legacy financial systems to SAP was driven by the desire of MIT's central administration to have a clear and unobstructed view of overall financial operations at the Institute. Subunits (departments, labs, centers) preferred systems that suited their local operations and goals. While there was considerable overlap between the goals of central administration and those of the subunits, there were also points of difference, which sometimes became points of contention. Without question, human beings were organizing this technological world. But whose organizational plan would dominate? Put a little differently: Who is the "we" that organizes the technological world?

In implementing a complicated software system like SAP, many conflicts over organizing the technological world take the form of conflicts between customization and standardization. According to Executive Vice President John Curry, "eighty percent of the cost of implementing new software involves customizations." Even if this is a ballpark figure, it makes the point that the cost of maintaining institutional habits is high. MIT's central administration preferred standardization because customizing was expensive. In the words of one senior officer: "If the vendor's worldview matches yours, you're OK. If not, one or the other has to change." This choice was confronted repeatedly in the transition to SAP. Since the cost of adaptation is high, the bias is toward standardization. But the social assumptions about staffing that are built into SAP software are not those of a university staffing structure. That is when the software begins to feel like technological determinism. It feels as if the software is pushing you around, demanding that you change your ways to fit its ways.

A good example of this pushiness is the controversy that surrounded the SAP module for purchasing information. In any financial system, purchasing is a fundamental process. A university buys everything from routine office supplies like paper clips and desks to one-of-a-kind lab equipment to services ranging from routine print jobs to high-end consulting. You would think that this "business-to-business" process is similar in universities and in other businesses, but it is not; the social structure is different. In business, those who handle purchasing data are usually specialized financial administrators who can take the time to learn a powerful, flexible, complex system like SAP, and who can keep up with upgrades because this is what they are hired to do. In a university, many financial transactions are handled by amateurs—faculty members, research staff, postdoctoral fellows, graduate students. They may be smart and highly computer literate; however, their main responsibilities lie elsewhere, and they do not have specialized expertise in handling financial data.

The confidentiality of purchasing data is more of an issue for amateurs than for specialist financial administrators. In a business setting, the individuals who do purchasing have access to these data on a need-to-know basis, and their entire job rests on their ability to keep this information confidential in a well-defined organizational structure. At MIT, it is less clear what graduate students and postdoctoral fellows, who do much of the ordering, know or need to know about the overall management of a lab. They might use purchasing information to extract other (possibly sensitive) information about research work not related to their own part of the project. The prospect of opening up purchasing data to much wider scrutiny made many MIT researchers uneasy. Some of them expressed their reservations to the Management Reporting Team, which had primary responsibility under Reengineering for implementing MIT's conversion to SAP R/3. The team considered these objections but decided that the cost of customizing the software was too high. In its weekly report of February 11, 1997, the team recommended that MIT "use native SAP authorizations and change its institutional culture to accept broad visibility of purchasing data." This language is an interesting mix of anthropology and change management ideology: the "native" rhetoric of SAP should force a "change" in MIT's "institutional culture."

The team's recommendation was met with an outcry from the departments, labs, and centers. They felt their way of doing things was the "native" one, and that SAP missionaries should change *their* ways. The reaction was so strong that the senior administration asked the Management Reporting Team to review the options. Its members did so and confirmed their earlier recommendation. Continuing to shield these data, they explained, would require either a separate authorization structure outside SAP or a significant modification to SAP: "Either of these shielding alternatives would have required a great deal of system maintenance, removed an important SAP functionality, carried a risk of incompatibility with some future versions of the SAP software, and increased the cost of the project. Furthermore, since SAP gives users the ability to get at certain data from a wide range of paths, we would never be sure that we had restricted access to every possible path to purchasing data."[22]

The administrators most concerned about the confidentiality of purchasing information agreed to survey colleagues at other universities where this information was more visible. Reports from their colleagues convinced them that a compromise was possible. MIT could move to the SAP authorization system, with unrestricted access to purchase-order data, so long as access to accounting statements continued to be restricted. Three years later, the purchasing system is operating without reports of serious security problems. On the other hand, some subunits have had to hire financial administrators to handle tasks such as purchasing, because SAP has proved too cumbersome for amateurs. This looks a lot like technological determinism: because the software is so complicated, a lab is forced to hire a financial administrator. But other people, outside the lab, chose that software in the first place to cut overall costs and to achieve overall clarity. In the case of purchasing information, if the financial cost of changing the software had been less the decision might have gone the other way. It also might have gone the other way if the cultural implications of changing purchasing procedures had been greater. But when the costs of customization were so high, the threshold for approving customization was also high. Those who wanted customization may win this episode or that, but the pressures for standardization are unrelenting, and so the losses keep piling up.

Clarity permits control. Administrators like to shine light in dark corners. One lab head said: "We need a budget in a form where we can *see* things. We need direct insight. Right now we are imagining things." Of course, he wanted a budget in which he could understand what was going on, but not necessarily those above him. When each department or center had its own ways of doing accounting, it could move funds around in ways that made it difficult for the central administration to figure out what was going on, and that allowed the head of a subunit to maintain control of as many funds as possible.

With SAP, as one senior administrator said, "They can't hide anymore. We can just push a button and get all those numbers." This would not necessarily mean that the central administration would get the dollars, but at least they would know better where those dollars were. Just because technologies are constructed by humans does not mean they do not exert control over human life. In many cases, they are constructed precisely to exert such control.

**Cultural Resistance**

As the Reengineering Project widened into a more ongoing and pervasive process of "change management," the justification for new software systems became focused less on direct cost savings and more on the need for integration. Since MIT's administration had invested so much in SAP software for accounting, it had an understandable bias toward leveraging that experience in using SAP systems for other purposes, such as payroll or student services operations. Integration became the new technological imperative.[23]

As MIT was busy implementing SAP financial software, SAP was busy developing new products for business and university

markets. One of these products was a student services information system, which handles registration, enrollment, grades,
degree audits, and the like. This SAP product was at a very
early stage of development in the late 1990s, and other student
information systems recently put on the market were in visible
trouble. Fortunately, MIT did not have an immediate need to
replace its student information systems. In the early 1990s,
MIT had built a system based on a commercial product (Ban-
ner) but heavily modified it so that it became a separate
species known as MITSIS (MIT Student Information System).
Few programmers could deal with its complexities, however,
and there was no outside market for MITSIS specialists.
Upgrading the system became progressively more expensive.
On the other hand, MITSIS was well suited to the unusual,
complex curriculum of MIT, full of quirks that represent the
outcome of educational compromises hammered out over the
years. I once remarked to Executive Vice President John Curry
that at MIT we have reified our complexity in our software.
Curry teased me: he knew he was at MIT, he joked, when he
heard someone use the word 'reified'.

In plainer language, we were back to the old tug of war
between customization and standardization. If you write code
as subtle and complicated as MIT's curriculum, you end up
with a map nearly as large as the territory. Writing and maintaining such code is inherently expensive. Yet the curriculum is
the educational heart of the institution, and most faculty members would strongly defend its details. It is one thing to modify
MIT's practices in purchasing data. It is quite another to
propose that MIT modify its academic program to suit an
off-the-shelf software system.

Because SAP was at an early stage in developing its student
services software package, staff in the registrar's office explored

the possibility of working with SAP in developing the new software. For a reasonable investment on our part (mainly the cost of backfilling staff who attended SAP workshops and meetings on the student services package), we might end up with a product that suited our needs. This would be a conscious project in what historians call the social construction of technology: getting involved with design at an early stage, in order to shape a technological world that we could live in. It might not be participatory democracy, but it was something like participatory capitalism.

Within a year, it was clear that this experiment was not going to work. The first problem was the shortage of staff time, which had been an issue in the Reengineering Project: how do you find time to plan the future when you are staffed only to maintain the present? Reengineering had provided resources for serious "backfilling," but Reengineering was over. In addition, the very reasons that made some MIT staff especially valuable to SAP developers—their detailed knowledge of the student information system—also made them especially valuable to MIT. They could not easily be backfilled.

The decisive problem, however, was that MIT's participation, no matter how vigorous, would not make much of a difference in the final product. SAP was developing software for the general university market, and MIT curricular rules are not readily transferable to other institutions. To the rest of the world, our features look like bugs. No matter what theoretical liberty MIT had to influence the writing of the code, in practical terms it had little influence because MIT was so unrepresentative of the market. The wider the market for the product, the more generic the product must be. MIT, with its non-generic academic rules, could change the rules to fit the market, or dramatically modify software bought on the market, or develop its own product at its own expense. We were faced not so much with technological determinism as with market determinism reified in a technological product.

Any form of determinism is less a command than a process. Evolving or replacing MITSIS, whenever that happens, will be a long and difficult process. The faculty will probably be asked to modify some of its rules to make software development easier, and the faculty will probably resist doing this. Technological change forces a series of decisions about money and control. People and institutions adapt, but adaptation is neither automatic nor straightforward. Tempers rise, compromises are hammered out, conflicts are mediated, tempers cool, and life goes on, not the same but not wildly different.

What is sadly predictable is that these conflicts will usually be described as ones where "culture" is "resisting" the "technology." In the language of change management, technology equals change, and culture equals resistance to change. This analysis is both simplistic and untrue, and its prevalence is one of the most unattractive features of Reengineering and similar change-management processes. People at MIT who "resisted" Reengineering were often said to be doing so because they wanted to maintain the existing culture, as if this were a fault. Because change is inevitable, it was further said, resistance was futile. Worse yet, it could border on treason. I heard some faculty members and staff described as "trying to stab Reengineering in the back," when they were only trying to defend an institution they loved. At one meeting, a presenter listed what she called a "risk analysis of change management," and one item on the list was the "friction of those who resist." The job of change management was frequently described as "getting people on board." This is why it took no little courage to speak out against Reengineering.

The analysis of "culture resisting technology" is itself irresistible to change agents. 'Technology' sounds hard and rational, while 'culture' sounds soft and irrational. This language has the effect of making the change agents into tough realists and the resisters into soft sentimentalists. It also has the effect—maddening, from the victim's point of view—of making a rational response impossible. When resistance is equated with cultural identity, discussion shifts from the objective level (Is this procedure better than that procedure?) to the subjective level (Why are you so hostile?). This therapeutic discourse automatically makes the so-called resister the loser. That person is labeled defensive and then, of course, cannot defend herself against the label without being even more defensive. Her feelings, rather than the technical choices, become the issue.

Since hearing these pointless and painful arguments, I have become much more careful, as a historian, in using the word 'culture'. As a historian, I had been used to thinking of 'culture' as a synonym for social context, the softer human elements around the technological hard core. But since technology *is* society, and machines *are* organization, the layered model must be discarded. Culture relates to technology not as surrounding context but as part of the design, in the form of tradeoffs. In the Reengineering Project and beyond, we were always making cost-benefit analyses to measure cultural change against software change. You could put them on the same spreadsheet, so to speak. Whether you stick with an existing system or go with a new one, you are always dealing with techno-cultural systems. Cultural work is necessary to support technical work, and cultural work also requires substantial investments in budget, space, and staff.

Though advocates of Reengineering liked to wrap themselves in the garments of tough technological inevitability, a large part of their work was cultural. They were determined to introduce a new culture to MIT: the genre of lists, the terminology of students as consumers, the definition of employees as performers. Most of the resistance to Reengineering was not to SAP or to any other particular technology but to this culture, perceived as alien and destructive.

In a recent collection of studies of so-called technological resistance, Martin Bauer observes that "the common-sense assumption" that "the object of resistance *is* technology" is problematic. "What is being resisted is normally complex and requires empirical analysis. . . . It may be useful to distinguish resisting hardware [or software, I would add] from resisting its consequences. . . . Often it is neither the design nor its consequences that are resisted, but the process by which the technology is put to work that is found wanting."[24] All these—design, consequences, and process—aroused resistance in Reengineering, but process most of all. Conversely, those who resisted often did so as much for "technological" as for "cultural" reasons. In student services, the staff are not anti-technology, much less machine breakers. Most of them love information technology. If they criticized technological change, it was not because it was technological, not because it was change, but because it seemed, in their word, "stupid." They felt they knew the technical issues in a detailed way, much better than any consultant, and they did not want an information system that was not well suited to their work. They felt they were the hard ones, the rationalists, the realists, the specialists who knew how to get the work done, not soft-skilled utopians interested in performance and role playing.

This is one of the oldest patterns of conflict in the history of technology: conflict between work performed with local knowledge and specialized skills and standardized systems introduced to enlarge markets, cut costs, and assert managerial control across space. Historians have studied this clash in widely disparate settings, ranging from cheese making in medieval Europe to machine tool making in the nineteenth century to management in the postwar United States.[25] Today the imposition of such standardized systems comes bundled with a set of cultural assumptions, which typically include the ethos of the marketplace, a belief in technological solutions, and a bias against unquantifiable realities (notably psychological and social ones).[26] These cultural assumptions are often invisible to their proponents, who believe that they are engaged in entirely rational analysis. Their systems are techno-cultural, but their own culture is invisible to them. When such systems are challenged, society is doing the work it should do. It reasserts reality in the face of abstractions. It adjusts the rate of technological change to fit the needs of human beings. It maintains institutions that help individuals adjust to change. The ideology of change management sees this work as negative, when in fact it is essential to maintaining the creative capacity of individuals and societies.[27]

**Blinded by the Light**

In a reflexive world, the "society" that is doing this necessary work is at the same time being transformed by the very technologies it is seeking to manage. For this reason, the conflict between local knowledge and large systems is more complicated than it was in the Middle Ages, or in the nineteenth century, or even in the immediate postwar decades. Information technology, employed on a large scale, has changed the way society manages itself on levels that are pre-political, subtle, and pervasive. These changes revolve around the social decision to "go digital," which long preceded SAP or Reengineering. If there is technological momentum today, it arises from this practice of reifying society in binary code. Ultimately, yes, it was human beings who decided to make this a common practice—but the decision point is utterly diffuse and long past.

The consequences of "going digital" seep everywhere, for it reflexively forces society to a level of enlightenment that has never before been sustained. I mean "enlightenment" not as a metaphor for rationality but in a more (though not entirely) literal sense. The explicitness of binary coding demands clarity even when individuals or organizations might prefer to do without it. "With digital, doubt no longer exists."[28] But human life is full of doubt and ambiguity. Thus the scene is set for ongoing conflict between the shadowy corners of life and the probing light of software that refuses to accept doubt. As software systems extend to cover more and more of human life, it is becoming harder to shield any part of society from the glare of visibility.

The risk of over-enlightenment may be seen in microcosm in MIT's Grouper Project, which began in late 1999. At that time, as part of its general activities in updating academic software, MIT's Office of Information Services launched this project to provide a comprehensive system for managing class lists. Managing class lists is as much a core business process for a university as managing purchasing data. Universities need class information to track registration, credits earned, and grades. They also need class lists to track revenue, to be sure that students have paid for the courses in which they are enrolled.

The Grouper team began to work on some undeniably useful software. For example, it produced a program that automatically generates a list of students for each class, based on information in the Registrar's database, that can be updated each night. The team started working on ways to link the class

list to the distribution of electronic materials to enrolled
students. One of the benefits of "going digital" is that both
teachers and students can have online access to class materials
and assignments. With this access, teachers can communicate
much more easily with students, and students with other
students.

On the other hand, privacy rules require that class lists
not be automatically distributed to everyone at MIT, much
less to the general public. Anything relating to an identifiable
student's participation in a class is part of that student's educational record, which federal law clearly defines as private.
Class email lists must be managed to prevent unauthorized
participation. For all these reasons, defining who is a member
of the class group is a more significant process in the digital
age than it was in the pre-digital age.[29]

Before long, the people working on Grouper were caught in
a cyberthicket. They found that it is not at all obvious who is
an official member of an MIT class. It is hard enough to define
who is a student at MIT, but at least for this step registrarial
and financial information provides basic guidance. Defining
a class enrollee is more complicated. The Grouper team discovered that some professors would allow or even ask a colleague, or a student who was not officially enrolled, to sit in
on a class on an informal basis. They were not official auditors
but unofficial participants. Such arrangements were not in
strict accordance with MIT rules, which required official student status for class participation. However, no one minded
when the practice was infrequent and when it apparently
caused no problems in the classroom. University life requires
judgment calls all the time. One of the strongest threads in its
social fabric is the understanding that faculty members have a
lot of discretion in making those calls.

Computers do not like judgment calls, however. In order to draw up class lists, MIT Information Services needed unambiguous answers. If these informal visitors were going to see class materials, they had to be declared students. If they were not declared students, they could not see the materials and therefore could not participate in essential class activities. They could not remain ambiguous. The Grouper team had long discussions about the distinctions among students, members of the class, and members of the community. This led to more discussions about the definition of a "person" at MIT. It turned out that there were three definitions of a full-time "person." One member of the Grouper Project proposed, not entirely in jest, that one was a person at MIT if one was a "Kerberos principal"— that is, if one had a username identifiable on the campus-wide computer system.

These issues, framed in technical language, did not get resolved. According to one leader of the Grouper Project, staff members working on it thought that the ambiguity of the physical analog was something that they could resolve in digital form if only they did enough clever programming. As time went by, they realized that the problem was social rather than technical. The Grouper Project summarized its conclusions in a memo that ended with a suggestion for a "General Discussion Topic":

As we move from a mostly analog world in the classroom to an increasingly digital world, some aspects of the physical world need to become more explicit. By becoming more explicit, we often shine a light on inconsistencies between practices, behaviors, and policies. . . . When a Faculty member's "view" of the classroom comes into conflict with explicit MIT policy, or de facto MIT rules, what is the MIT philosophy for resolving such disputes?

This question shows an understanding that the issues go beyond policy and rules to underlying "philosophy." But how is MIT to develop such a "philosophy"? By its nature, human society is complex, messy, ambiguous, subtle. By its nature, computer code is structured, unambiguous, logical. It is not clear that any "philosophy" can reconcile these two realities; its role may be, rather, to analyze the conflict. Also, it is not clear to whom the question is addressed. Who is going to articulate the MIT "philosophy" on such matters? Who is MIT here?

MIT does not usually have difficulty articulating institutional policies. In particular, it has well-established processes for formulating policies about privacy. Toward the end of my time as dean, we completed a major review of privacy concerns related to student information in the digital age. But the privacy issues raised by "going digital" are larger than the distribution of electronic information or electronic snooping. In her book *In the Age of the Smart Machine*, Shoshana Zuboff uses the Panopticon—the model prison where prisoners are always exposed to the gaze of jailers, whether or not the jailers are actually looking—as a metaphor for the way computer technology expands the possibilities for surveillance of workers.[30] These possibilities may be alarming, but policies can be put in place to deal with them.

The example of the class list shows that computers do not have to be used for surveillance to threaten privacy. Software, any software, just by being digital, is potentially a Panopticon. Even in routine management operations, just by demanding clarity, software intrudes into large areas of human life that have been left murky. It is in the inherent nature of computer code to "shine a light on inconsistencies between practices, policies, and behaviors."

The paradox of information technology is that seemingly insubstantial machinery, constructed from light particles and digits, is so powerful in making visible social patterns. The drive to raise self-consciousness is one of the strongest cultural imperatives in information technology and management philosophy today. Reengineering demands that workers think critically, analyze, self-scrutinize, self-realize. In this new world, doing your job is no longer enough: you are supposed to hold yourself and your job up to the light, turn it over, examine it, rework it, and examine it again.[31] This utopia of self-actualization is also one of relentless self-scrutiny.

But how much clarity can individuals and organizations tolerate? For individuals, the role of the subconscious and routine in life is not going to disappear, nor should it. Limiting the glare of visibility is not the same thing as allowing the triumph of the dark side. While explicitness is functional for the computer, it can be dysfunctional in social life. Society is often well served by blurred edges, unclear definitions, diplomatic circumlocutions, and fuzzy logic. Social logic is different from computer logic, and the ability of society to do its work is damaged if it restricts the ability of human beings to do what they do best: be creative and use judgment.

### Technological Change vs. Historical Change

The Reengineering Project officially ended on June 30, 1999. The budget gap had been closed, but not through Reengineering. What happened at MIT in the later 1990s is what happened in the economy as a whole. The answer to financial problems came less through management techniques and accounting systems than through creative exploitation of new opportunities, executive leadership, and a raging bull market. Instead of dramatically lowering the slope of the top line on the budget chart, MIT significantly raised the slope of the bottom line. MIT embarked on an ambitious capital campaign and a series of industrial and other corporate partnerships. President Vest worked hard, with considerable success, to stabilize government support for research. His three-prong strategy proved a success.[32]

At MIT meetings, the chart showing diverging lines of spending and revenue was replaced by a much more heartening bar chart showing the changing sources of MIT's income. The chunk representing government sources kept shrinking in the 1990s, while the chunks representing corporate, individual, and other giving kept getting bigger. By 1998, U.S. industries were contributing $2.4 billion to the support of research universities, compared to $133 million in 1976. Private and corporate giving began to outstrip government support. In the 1960s, the federal government had provided roughly two-thirds of the national research and development funds, with private sources contributing one-third. At the end of the millennium, the ratio was two-thirds private and one-third federal.[33]

Reengineering played an important if indirect role in these changes. In and of itself, it did not harvest anything like the savings that had been hoped for. It had great value as a form of technological display, however. Reengineering demonstrated to the MIT Corporation, to potential donors, and to the government that the Institute was serious about improving efficiency and managerial oversight. The increased flow of funding from all these sources depended, in a significant though ultimately unquantifiable way, on their renewed trust in MIT financial management.

Within MIT, the legacy of Reengineering was much more ambiguous. It brought such accomplishments as the Student Services Center, increased staff mobility across the Institute, and useful new software systems for processes such as purchasing.

But if it had started with a bang, it departed with a whimper. The ideological hard edge of Reengineering became worn before its official end. Employees lost their acute fear that they would soon lose their jobs. They also lost any illusions that Reengineering would bring a new world of work. Reengineering was now identified with SAP, and talk of change was limited to the latest SAP upgrade.

Along with most of the staff, I was happy to see Reengineering come to an end. We also shared a sense of melancholy, of regret for promises unfulfilled. We remembered the early Reengineering meetings, at which we had imagined a new world of student services that would be an integral part of an MIT education, not a refuge from it. But we had not succeeded in creating a new world in which MIT would offer significantly more support for student educational experiences outside of the classroom. Eventually the latter happened, but not through Reengineering. Indeed, a new world of student services could not have been created through Reengineering. Its creation depended on a deep cultural transformation at MIT, from a culture that most of all prizes individualism and small living groups to one that puts equal value on social experiences and the larger community. The change was already underway; students, faculty members, and staff all understood, though in different ways and on different levels, that MIT's respect for student independence should not lead it to neglect student life. But such a change could not come about through teams and visioning sessions. It would require strong leadership and the involvement of the entire community.

The ideology of technological change implies that market forces have replaced historical forces. This is not so and will never be so. Historical change is not planned, nor is it avoidable. What drives it is more primal than teams or technology.

Historical change happens when people's deepest feelings get involved, when they are confronted by a crisis, when they have to question their habits and assumptions—even cherished ones, because sometimes habits and assumptions are no longer sustainable no matter how much they are cherished. This kind of change calls upon human beings to assume responsibility and to act courageously. It involves suffering, conflict, and sometimes death.

On September 30, 1997, Scott Krueger, an 18-year-old MIT freshman, died in a Boston hospital from acute alcohol poisoning. He had been at MIT for barely a month. During his very first week he had pledged Phi Gamma Delta, a fraternity located in Boston's Back Bay across the Charles River from the MIT campus. He and seventeen other pledges had moved into the Phi Gamma Delta house. According to the grand jury that investigated his death, in the middle of the last week of September, all the pledges were told they had to participate in an "Animal House Night" that Friday evening. After watching the movie *Animal House*, the pledges were introduced to their fraternity "big brothers," who presented their "little brothers" with gifts of alcohol. In the aftermath of this event, Krueger drank most of the bottle of Bacardi spiced rum that his big brother had given him. He passed out, was carried downstairs to his bedroom in the basement of the fraternity house, and choked on his own vomit. When his brothers came to check up on him, he had turned blue and was in a coma. Three days later his family took him off life support and took his body back with them to their home town of Orchard Park, New York.

The MIT community was stunned. We thought we looked out for one another. How could such a tragedy have happened here? In the months that followed, the community discussed student drinking, fraternities, and the housing system. All these issues were tightly coupled in the circumstances of Scott Krueger's death. MIT took both immediate and long-term steps to reduce student drinking, including the formation of a high-profile working group, co-chaired by Phillip Sharp (a Nobel laureate professor of biology) and Mark Goldstein (the head of pediatrics at MIT Medical). This group concluded that proposals to curb alcohol abuse should be linked with other structural changes, such as housing freshmen on campus and increasing the supervision of student residential and campus life. Many other groups and individuals in the MIT community echoed these conclusions. Still others challenged it.

In particular, the wisdom of allowing one-third of MIT's first-year students to live in fraternities (and a much smaller number in other independent living groups) began to be debated. Immediately after Scott Krueger's death, President Vest announced that a new undergraduate residence hall would soon be constructed. Would this additional capacity be used to require all freshmen to live in MIT-operated residences? Many students feared this would effectively end fraternity life at MIT, where it had flourished almost since its founding. Even more students feared that requiring freshmen to live in Institute residences was the beginning of the end of student choice in housing decisions. Many alumni and alumnae, especially those for whom fraternity or independent group living had been the best part of their MIT experience, were strongly opposed to requiring freshmen to live in MIT-operated residences.

In the year after Scott Krueger's death, the MIT community was consumed by sometimes ferocious debates over housing. Over them hovered the handsome, smiling image of Scott, whose senior yearbook photograph appeared repeatedly in the local and national media. So did a quote from two students who in the early 1990s had been active in investigating and publicizing their view of the problems in MIT's fraternities. In 1992 they wrote to President Vest to ask: "When a student is killed or dies at an MIT fraternity, how will MIT explain its full knowledge of dangerous and illegal practices persisting unchecked over a period of years?"

President Vest, I, and many others have anguished over this question. Eventually, MIT had to make significant changes in its housing system, because of the larger trends of history. The large and rapid increase in the number of women students at MIT, beginning in the 1970s, radically reduced the pool of potential pledges. Skyrocketing real estate values in Cambridge and Boston, especially in the neighborhood where the fraternities are concentrated, led to higher expectations for quiet and order and greater scrutiny by regulatory agencies. At the same time, fraternities found themselves financially squeezed by the artificially low price of MIT student housing. The quality of life in the fraternities progressively declined in the postwar years. Fraternities also found it harder and harder to afford liability insurance and to convince alumni to assume financial and legal responsibility for houses. But it took Scott Krueger's death to get beyond awareness to action, by creating a sense of urgency about finally addressing long-standing problems. It also took President Vest's courage to make the decision to require freshmen to live on campus, once he had determined that this was the right thing to do. The decision was announced at the end of August 1998, about eleven months after Krueger died. Vest's assumption of personal responsibility, both in making that decision and later in offering a public apology to the Krueger family, were critical steps in moving beyond the tragedy to meaningful change.[34]

Four years later, student life at MIT is a new world. Cranes bend over the construction site of the new residence hall.

In freshman orientation, residential choice receives less attention, while institutional norms and academic choices receive more attention. In the summer of 2002, entering freshmen will have the chance to select their residence before they get to MIT; many of them will know where they will live when they arrive on campus. Fraternity rush will take place later in the fall, and pledges will not move into fraternity houses until their second year at MIT. Enforcement of rules about underage drinking is tougher, not only on the part of MIT authorities but more generally in the cities of Boston and Cambridge. Students are more aware of liability issues, safety issues, and their responsibilities toward one another.

To be sure, the changes should not be overstated. Underage drinking and heavy drinking still occur at MIT, and too many students still arrive assuming that drinking will be a significant part of their college experience. But attitudes have evolved. The reverberations from Krueger's death continue to alter campus life, not through policy announcements but through the accumulation of individual reflections and choices. In the fall of 2000, when MIT's settlement with the Krueger family was announced, a student leader in the fraternity system recalled: "Scott's death made me want to make sure that none of my friends would ever die from alcohol again as long as I know them. That's why I became the IFC [Inter-Fraternity Council] Risk Manager in 1998." Social life at MIT, he said, has "changed in every way imaginable since 1997. . . . I think that the sacrifice of 'fun' pays homage to the memory of my close friend Scott and shows that some of us took everything positive that we could from his death."[35] For a student to use the language of penance and sacrifice is an extraordinary change from the rhetoric of individual choice that used to dominate nearly all discussions of student life. More and more one hears instead the language of "community." A former president of the Undergraduate Association commented: "Students have more of a bottom-up approach to building community. The administration works from the top down. You need both to make changes."[36]

The linking of change with community became more and more of a theme during the late 1990s. Both Reengineering and the new rules about alcohol and housing put enormous strains on the MIT community. In many ways, student resistance to the decision to house freshmen on campus echoed staff resistance to Reengineering. Students felt that an alien culture was being imposed on them, and they were resentful of being subjected to closer central control. If SAP software sometimes felt like a Panopticon, so did a new housing policy that required adult resident advisors in fraternities, sororities, and independent living groups. In both cases, MIT as an institution was asserting more central control, largely in response to pressures outside MIT—in the one case financial, in the other case regulatory. These "top-down" actions were necessary for MIT to continue to function as an institution. This story has been repeated over and over in recent years, in a range of operations. MIT has tightened central regulation in response to government requirements for closer monitoring of research funds, of the outside professional activities of faculty members, of laboratory safety measures, and of countless other risks. In many of these cases, faculty members object to the increased surveillance, which they regard as an unnecessary nuisance, but the institution has to conform to the regulatory environment.

In Ulrich Beck's analysis of "advanced" or "reflexive" modernity, "the social production of *wealth* is systematically accompanied by the social production of *risks*." Alcohol and fraternities are hardly new risks, but tolerance for their hazards is reduced in a society where the management of risk, as a collective rather than a personal responsibility, has assumed great practical importance. In a risk society, appeals to community norms get lost in the wind. Instead, the discovery, the administration, and the remediation of risk are handled systematically through well-defined rules and structures.[37]

Over the generations, people at MIT have developed a complicated system of self-understanding and self-regulation, embodied in language, relationships, social norms, and habits. We appeal to this system to explain ourselves to ourselves. It makes sense to us, as a community, but it does not make sense to regulatory agencies such as the Boston Licensing Board, or to the Middlesex County grand jury, or to the larger public. Like most communities, we were happier leaving things unsaid or half-said, unseen or half-seen, with much room for variety and flexibility within broad general expectations. Managers of risk society, on the other hand, demand enlightenment, specificity, and visibility.

In the tumultuous months after the death of Scott Krueger, producers from the ABC television news program *20/20* (whose very name reinforces the point about visibility) informed MIT that they were going to do a program on college drinking. They needed an on-camera spokesperson from MIT, and I was assigned this unenviable responsibility. To prepare, I got some media training. First question from the media trainer: Who is in charge of the MIT fraternities? There are many layers of regulation, I began, including the fraternity officers, the house corporation, the Inter-Fraternity Council, the dean's office, the national. . . . The media trainer cut me off. She shook her head and snapped: "That sounds like no one is accountable. It's much too complicated. The answer is that MIT is responsible and you are in charge. Let's start over again. Who is in charge of the MIT fraternities?" My media trainer prepared me well. As we taped MIT's part of the *20/20* show, the interviewer kept asking me: "Who is responsible for the death of Scott Krueger? Who's responsible?" There is no single party, I responded, but MIT certainly bears part of the responsibility—and ultimately MIT is in charge. When President Vest offered his personal apology to Mr. and Mrs. Krueger, he was repeating this message.

Being a community is not enough. In a risk society, institutions need clear accountability and lines of responsibility that are transparent to the world at large. But transparent structures of authority are also not enough. For truly effective regulation of heavy drinking, MIT also has to function as a community of mutual trust and concern.

In the *20/20* interview, I tried to explain how policing alone would not reduce student drinking, but might only drive it underground or off campus. Even in the best-run dormitory, I said, a university cannot prevent a student from obtaining alcohol, especially in an urban setting, or from drinking that alcohol in his or her own room, or from inviting a few friends over to join in. MIT cannot run a 24/7 surveillance system, I said. Even as a parent, I cannot do this; at some point I have to trust my children to do the right thing, and at some point a university has to trust its students. If community pressure contributes to alcohol abuse, it also can be an effective check on alcohol abuse. ("Scott's death made me want to make sure that none of my friends would ever die from alcohol again. . . .")

We had an institutional dilemma. Just when we needed the strength of community norms to regulate student drinking, our sense of cohesion and trust as a community had been undermined by repeated efforts to exert more institutional control. Closing the financial gap and the regulatory gap had increased the trust gap. Students worried that "the administration" would punish them for trivial offenses. They worried that calling for medical help for a drunken student, for example, could lead to getting kicked out of MIT. The more the trust gap widened, the more we talked about the need to "build community." How was MIT to balance its need for trust and its need for management?

In the words of one vice president: "We used to be a community, and now we are becoming a corporation." A corporation may seek to reduce risk and manage change, but its techniques may undercut the very social structures that help people make individual decisions about reducing risk and adapting to change. MIT has to be both a corporation and a community. How can this be accomplished in the information age?

# 4

## Technology and Community

### Change Talk and Community Talk

Every MIT undergraduate residence and most graduate ones have a faculty housemaster who lives in the building (often with spouse and children) and who provides oversight of its intellectual and social life. It is one of the most difficult, important, and rewarding roles that faculty members play at MIT. Each year, President and Mrs. Vest invite the housemasters to dinner to thank them for their service and to hear, in a relatively informal setting, how things are going. In the spring of 1995 the after-dinner conversation turned to a series of recent events in the residences. None of them was major, but their integrated effect was to make the housemasters aware that they did not clearly understand the educational principles underlying MIT's residential system. They asked Vest to help in clarifying those principles.

The day of a university president is spent listening to people with concerns, but Vest felt that this group had raised an especially valid point. The MIT faculty had last done a comprehensive review of both residential and academic education almost half a century earlier, at the time of the Lewis Committee. MIT was then still largely a commuter school. Some of the students who did not live at home found lodging in a few student residences on the Cambridge campus, where the Institute had moved in 1916; many more lived in fraternity houses, most of them (then as now) located across the Charles River in Boston. The Lewis Committee formed a subcommittee that recommended a substantial institutional commitment to developing a true residential campus, so that the informal education provided by student life would become a part of MIT's overall educational experience. After some thought, President Vest concluded it was time to undertake another high-level review of MIT's educational mission, with special emphasis on the relationship between campus experience and academic experience. When he and I discussed the dean's position in June 1995, he told me that he was planning to form a Task Force on Student Life and Learning on the model of the Lewis Committee.

Beginning in the fall of 1995, in collaboration with the president, the chair of the faculty, and dean's office staff, I began to recruit faculty members to serve on the new task force. The first criterion was a reputation as a "good citizen"—we were looking for individuals with broad interests who regularly allotted some time to Institute-wide commitments. Even for a group of "good citizens," joining the task force was far beyond the normal call of duty. Setting aside at least a year for regular meetings was an extraordinary commitment. No one imagined it would take two years.

For the average faculty member, a personal cost-benefit analysis works against committee service. Unlike research and teaching, it does not count directly toward salaries and promotions. In fact, some department heads regard too much committee service as a negative sign. Most faculty members regard committee service as the academic equivalent of jury duty—a responsibility that one would prefer to avoid, though it is necessary for the larger good of the institution. Just because committees are so necessary, there are a lot of them: search committees, promotion committees, curricular committees, prize committees, policy committees, and many others, on all levels (departmental, school, Institute-wide). And just because there are so many committees, it is easy to forget how vital they are in running the Institute. If the MIT Corporation (the equivalent of a board of trustees) governs the Institute from the top down, committees of the faculty govern it from the bottom up. For every faculty member, there is an ongoing tension between the reward system for personal achievement and the reward system for participation in the maintenance of the institutional commons.

The faculty members who agreed to serve on the Task Force on Student Life and Learning ultimately did so because they were convinced that the task force's work would lead to serious, substantial change. The common phrase they used to explain their motives was "wanting to make a difference." This significant phrase is a reminder of the ancient origins of democracy, in the desire of mortal individuals to achieve an immortal glory, through shared words and deeds that are passed down as organized remembrance.[1] "Making a difference" is a phrase that suggests a motivation that is neither individual self-interest nor altruistic communal interest. In joining the task force, faculty members expected to exert political power and even to go down in institutional history, by having their names attached to a significant set of recommendations. (As one of the members of the task force noted, MIT still talks of "the Lewis Committee": no one knows it by its official title, the Committee on Educational Survey.)

By the early spring of 1996, co-chairs for the task force had been chosen: Robert Silbey from the Department of Chemistry and John Hansman from the Department of Aeronautics and Astronautics. The decision to have one co-chair from science and one from engineering was a deliberate effort to balance the interests and outlooks of the two dominant forces in MIT academic life.[2] Gradually, seven other faculty members, including myself, were added to the group. The co-chairs began the process of selecting student members, two undergraduates and one graduate, in collaboration with student government procedures. Finally, in July 1996, President Vest announced the appointment of a Presidential Task Force on Student Life and Learning to "undertake a comprehensive review of the Institute's educational mission and its implementation."[3]

For the next two years, the task force and Student Services Reengineering were the two poles of my existence, as they were of MIT's institutional life. In both we heard a lot of "change talk." In Reengineering this referred to supposedly sweeping, irresistible technological change that would usher in a new organizational world—a market-oriented workplace of self-empowered individuals who fearlessly promoted yet more change. When applied specifically to university organizations, technological change talk focused on the promise (also a threat) of "distance education." In the age of the Internet (the argument went), colleges and universities no longer have a near monopoly on valuable information sources. Since the price of gathering people around those sources at the same time is so high, market determinism combined with technological determinism would work against residential education.[4] Another dean remarked to me: "Everyone is scared that we have built a field and they will no longer come." Irresistible technological change might doom the residential campus.

You would think that the educational version of change talk would be welcomed at future-oriented, technology-oriented

MIT. You would think that MIT faculty members and students would love the idea of leaving behind the academic city on the hill for a globally reaching, technologically nifty educational model. Most faculty members, including those on the task force, have a large number of commitments outside MIT and lead neo-nomadic lives—which is why assembling the task force for regular meetings was such a headache for two years. Most students aspire to live similar lives.

But this is not at all what happened. The task force, a group of future-oriented technophile, students and faculty members alike—came to the conclusion, strongly and unanimously, that the future of MIT depends not on promoting technological change but on maintaining a beloved community[5] in Cambridge, Massachusetts 02139.

Those of us who served on the task force rejected the view of history as "technological change" bearing down relentlessly on doomed "communities." We challenged the assumptions that the new world of systems is the source of future vitality and that communities are the source of resistance based on the past. In the process of working through the issues of student life and learning, we concluded that locally and culturally defined groups are often sources of change—meaningful change, not the incremental variety. Over two years of study, we concluded that MIT has been so fertile in technological creativity—so innovative, if you will—because of informal groups, unstructured encounters, odd connections, wandering, and daydreaming. MIT's common resources gave us time and space to graze. If the technological systems of the information age enclosed the commons, we were in trouble. We defended turf against information technology.

This is an entirely counterintuitive outcome. This is MIT? It is MIT, in the way that the task force rejected anti-technology sentimentalism. The task force broke two conventional links: that between change and realism, and that between community and romanticism. In observing the process of Reengineering, which was going on at the same time, task force members concluded that the business-oriented "change agents" were the romantics. Reengineering advocates might talk in tough neo-realist tones about getting out of the line of fire of technological change, but their description of that change featured neo-romantic, even utopian promises about a brave new world of innovation. Instead of utopian fantasies of "making work go away," the task force presented more realistic assessments of resources needed to support creativity and discovery. In our conversations about educational community, for better or for worse, we on the task force did not have in mind the images of green playing fields, ivy-covered walls, and long conversations about the meaning of life. In a quite hard-headed way, we argued that the sources of creativity necessary to engender change, technological or otherwise, flourish only in a setting with time and space for the intense social interactions that are at the heart of both research and learning.

As pseudo-political discourses, change talk and community talk coexist everywhere today, rarely as dialogue, mostly as separate discourses that run on without many points of intersection. They represent different ways of thinking about the governance of technology. Change talk emphasizes participatory management, but this participatory ideal is constantly and decisively undercut by the fact that most employees have little real job security. They may be asked to give input to management; however, they are still the ones who are being managed and whose jobs are at stake in any "downsizing," and so their input is rarely fearlessly honest.

The tenured faculty "good citizens" of the task force, on the other hand, whose jobs were not at stake, could engage in something more nearly resembling participatory democracy. Working from relatively safe ground, the task force sought and received the input of other stakeholders—untenured faculty members, staff, and students. Together, gradually and cautiously, we became engaged in Latour's definition of Politics: the "progressive composition of a common world"— or, to use a term with an honorable history in Massachusetts, the composition of a commonwealth.

While the task force was doing this work, however, we were still part of the existing world and therefore part of the problem we were addressing. In this circular and reflexive world, how can you reassert the importance of community life unless you already have a reasonably strong community life? Slowly and carefully, we managed to construct a shared view of MIT that transcended our individual subjective views of the place.[6] How we did this—the process by which we arrived at our recommendations—is as interesting as the recommendations themselves. The story of the task force recapitulates and demonstrates the process of constructing a common world in the information age.

## The Task Force on Student Life and Learning

On August 1, 1996, eight of the nine faculty members of the Task Force on Student Life and Learning met for the first time in the Carriage House on the grounds of the Woods Hole Oceanographic Institution on Cape Cod. (Student members were not chosen until the next month, when students returned to campus after the summer break.) We had hoped for sunny skies, but the rain pounded relentlessly. Despite its grand name, the Carriage House was basically a cottage, and the conference room where we met was cramped and dark. After a year of trying to get the task force launched, I thought I would feel exhilarated. Instead, I felt damp and vaguely depressed.

We spent the first morning introducing ourselves to one another, explaining how we had become interested in what we do, how we had come to MIT, and how we had experienced MIT. We quickly discovered that our experiences overlapped very little. No matter where we had ended up in the institutional structure, no matter how much general service we had performed, none of us had ever lost the angle of vision from which we had first encountered MIT. We taught different subjects; we belonged to different professional groups; we inhabited different parts of the campus. At times it hardly seemed we were at the same institution.

We also shared little in the way of public history. This had not been a problem for the members of the Lewis Committee. For them, World War II provided a common point of reference, a pole star of first magnitude. Fifty years later, in our opening self-introductions, we found few common points of historical orientation. Those we shared dated back to the 1960s. Most of us had been undergraduate or graduate students in those years, and, whether up close or at a distance, most of us had been involved in civil rights or antiwar activities. In more recent years, our common history was thin. Global markets, global warming, and the information revolution were processes, not events; we had all experienced them, but in disparate and inconclusive ways. Historical processes do not have the binding power of historical events. If anything, they scatter people, by affecting them in such very different ways. In short order we realized that we were going to have to create a common reality, because we did not start with one.

Much of the first morning was devoted to venting about MIT experiences that are for the faculty the usual suspects in the decline of academic civilization: admissions standards (never high enough), the freshman year (no one in charge), laboratory experience (more needed), the introductory physics course known as 8.01 (why did so many students fail?), humanities (too disconnected), student motivation (never strong enough). By midday we were experiencing what one member likened to "8.01 drag." We would start talking about the larger changes in the world—global environmental problems, the disparity between the developed and the underdeveloped economies, changes in student demographics—but just as we began to rise to a loftier view, we would descend back to discussions of why students did not do well in 8.01, as if it would endure forever as a feature of the MIT landscape. We hardly discussed student life outside the classroom. As faculty members, we tended to equate education with academics. The rain droned on.

But the MIT ethos is that "we solve problems." One of the best qualities of this institutional culture is the habit of taking on a huge, apparently intractable problem, refusing to let it overwhelm you, and doing your best to break it down into manageable chunks. We reviewed our charge and began to plan how we would respond to it. John Hansman reminded us that we were "in an information-gathering mode." We started to plan the information-gathering process: reviewing data from surveys, conducting further surveys, obtaining documents, interviewing people at MIT and outside, preparing questions for those interviews, assigning research topics. Last but hardly least, we pulled out our calendars to find times when we could meet for two hours each week to gather and discuss all this information.

In the information age, nothing is less obvious than what information is and how it can be turned into shared knowledge useful for composing a common world. We decided to cast our net with deliberate non-selectivity. We decided to be deliberately inefficient, because all we knew at that point was that we did not know enough even to define the problems. If we narrowed our questions too soon, we might never ask the most important ones. Also, in an uncynical way, we realized that the credibility of our information was at least as critical as its functional role. We wanted to be able to defend our conclusions by going through a process of consultation that was as open and broad as possible. In this respect, we were like the Reengineering Project: we understood the display value of going through an intensive process, no matter what its outcome.

With the beginning of the fall term of 1996, the three student members were named, and we began our rhythm of two-hour weekly meetings. We started by interviewing the dean of admissions, the chair of the faculty, the executive vice president, the director of financial aid, the head of dining services, a group of housemasters, and the president.

Meetings are as much a part of the information age as computers. Meetings are the social practice that transforms facts into meaning. Outwardly, the meetings of the task force resembled those of Reengineering teams: coffee, juice, platters of melon, grapes, pineapple, and assorted cookies, a big table, committee members seated (sometimes sprawled) around it, the guest or guests at one end. The resemblance was superficial, however. Reengineering meetings were designed for Steering Committee members to hear reports about projects; we were asked for approval or advice, but there were few opportunities for genuine discussion and intervention, and so the Steering Committee too often failed to steer.

Task force meetings, in contrast, were more often conversations, not presentations, and hybrid conversations at that. We had a list of questions for each guest, but we also recognized that some of the most important observations might come from odd and unsolicited angles. Not surprisingly, task force meetings tended to ramble, sometimes badly. On more than one occasion, I found myself wishing for the brisker pace and greater discipline of the best Reengineering sessions. Our conversations had some wonderful moments, but those moments rarely connected. As the fall semester drew to a close, the significance of our fact finding seemed to be eluding us. The information map was beginning to feel like an information maze. We were still having trouble rising above local forces and getting to big ones that might transform all of MIT. Even more, were we having trouble rising to the level of forces for change affecting the world beyond MIT.

We decided that we ought to make a special effort to gather information from MIT's junior faculty—the younger, untenured professors who represent the future of the Institute, and who also represent a spectacular assemblage of talent. We decided to invite the junior faculty to a workshop in January, during MIT's long break between fall and spring semesters. There was quite a bit of give and take concerning the format of the workshop. Some of us wanted an all-day event, reasoning that anything shorter would not give the junior faculty members time to provide useful input. Some worried that this was asking much too much of these faculty members, who—because they are up for tenure, and because many of them have young families—are arguably in the busiest and most stressful period of their lives.

With some apprehension, we decided on a full-day workshop. To our amazement, 75 junior faculty members showed up.

Evidently they too wanted to enjoy collegiality, discuss high-level issues, and maybe make a difference. Many of them had come to MIT from other institutions, and they had not been at MIT long enough to suffer from 8.01 drag. By showing up in such numbers, and by investing the day with enthusiasm and creativity, the junior faculty reoriented the task force toward a comprehensive and expansive view of MIT's educational mission. The junior faculty workshop turned out to be the most stimulating of all our information-gathering activities.

From that point on, the information-gathering process began to open up to more and more non-faculty members of the MIT community. While some members of the task force continued interviewing administrators, the co-chairs began some parallel processing by establishing two subgroups: a Student Life Subcommittee and a Student Advisory Group. The Student Life Subcommittee was dominated by dean's office staff members, people who worked directly and consistently with students. The Student Advisory Group was made up of about two dozen self-selected graduate and undergraduate students, including the three students who were official members of the task force.

The formation of the two subgroups led to an unforeseen shift in the dynamics of the task force. An "information-gathering mode" may sound like a neutral process, but it is not. Someone is going to define, implicitly if not explicitly, what information matters. In the task force's early meetings, faculty members shaped the process by focusing on curricular issues: what to do with 8.01, how to balance science and engineering, how to get more flexibility and hands-on learning into the curriculum. What had begun as a faculty-centered process, however, evolved into one that was much more evenly balanced among faculty members, staff, and students. Many of the faculty members on the task force had not had sustained interaction with student life deans (the staff who worked most directly with students on many aspects of residential and campus life). When faculty began interacting regularly with the Student Life Subcommittee, they deepened their respect for the knowledge and dedication of the staff. I clearly remember the moment in one meeting when a faculty member of the task force first referred to the staff, with evident deliberateness, as "professionals."[7]

Even more striking was the change in social dynamics when the Student Advisory Group got up and running. As students gained more of a voice in the task force, they redefined the issues. Instead of being represented through the eyes of the faculty, they represented themselves. Instead of being faces in an 8.01 lecture hall, or even in a humanities classroom, they became visible as young adults whose lives, while dominated by classes and homework, also included roommates, hall mates, friends, lovers, meals, sports, parties, worship, families, and dreams for the future. Staff and students began to focus the attention of the task force on non-academic issues: counseling, advising, dining, housing, athletics, social life. The shift was not a matter of getting more or even better information. It was a shift in power that led to a shift in defining what information really mattered.

In February 1997, a month after the junior faculty workshop, the Student Advisory Group met with the task force and floated the idea that the task force endorse a "triad" of educational activities at MIT: student activities, academics, and research. The latter two were well understood as educational activities: the first was not, but the full task force was, on the whole, supportive of the concept. With its encouragement, the Student Advisory Group drafted a preliminary report refining the concept of the "educational triad." In this draft, written in the summer of 1997, the student authors renamed the critical third leg, the one that would be a new addition to academics and research: instead of student activities, it became "community." The students had decided that "student activities" was too narrow a term, too easily interpreted in a banal sense of joining a club. It also failed to convey the essence of what they most wanted: more sustained interest on the part of other adults on campus in their full lives, not just their academic lives.

As we discussed "community" in the task force, we continued to push for a more inclusive definition: we tried to articulate the goal of integrating the lives of faculty members, students, and staff. We also talked more about the importance of the campus as the geographical site of this human integration. In one informal conversation, a student remarked that what he missed most at MIT was "a sense of place."[8] This struck a resonant chord with many of us, at a time when a lot of institutional attention was focused on partnerships with outside corporations and universities, and when many of us were spending more and more time on commitments and activities and travel beyond MIT. Despite these trends, or more likely because of them, we began to talk about the need for the task force to endorse a "Cambridge 02139 philosophy."

While the students were developing the concept of an educational triad, faculty members worked on that of a "new quadrivium," a new type of liberal education that would include the natural sciences, the arts and humanities, the social sciences, and technology. These curricular ideas did not get very far, however. The initial disparities in outlook and experience so evident at our August 1996 retreat may have been moderated through our year of common effort, but they were still present just below the surface. The engineering view of the purpose of undergraduate education—i.e., developing competence or at least semi-competence in a professional track—was still far different from the arts-and-sciences view of undergraduate education as a general intellectual foundation for lifelong education. These differences have a long history and involve structures of professional development that extend far beyond MIT. They did not disappear just because task force members came to know one another much better and to appreciate one another's experiences more fully.[9]

We made some headway when, instead of listing what should go into the curriculum, we decided to articulate what we thought should be its educational results. We began to discuss what should be the "characteristics of an educated individual" at the beginning of the twenty-first century. We could agree more on the ends of a technically oriented education than on the means. Still, this question pushed us to confront some underlying differences. When one faculty member from engineering proposed that an educated individual today is familiar with "the integration, synthesis, and decomposition of systems," a scientist on the task force objected that this language threw the list off kilter because it smacked too strongly of engineering. He suggested something like "has the ability to manage complexity," which was accepted. Although we never did agree on a "new liberal education" or a "new quadrivium," our description of the "attributes of an educated individual" had a similar effect in expanding the significance of an MIT education well beyond its technical content.[10]

Early in the fall term of 1997, as task force members were exchanging draft reports, Scott Krueger died. The Institute was swept into anguished discussions about alcohol, student social life, housing, individual choice, and institutional responsibility, and the task force was swept along with it. While we talked about all these issues at length, the co-chairs wisely kept us from getting diverted into detailed policy decisions. For our report to be of lasting value, they argued, we had to take a lofty view, one above the pressures of the moment. Nevertheless, we could not ignore our strong differences, particularly on the critical issue of freshman housing policy. Faculty members, for the most part, felt that freshmen should live in dorms. Student members, for the most part, felt they should not have to.

We had agreed on the general principle of community; now we were made aware of the vagueness of this term. For most students, the paradigm of the MIT community was their living group, whether an Institute-run residence hall or a student-run independent living group. For most faculty members, advocating community for MIT meant something more academic in orientation and more comprehensive in scope. In the normal course of life, vagueness can be useful. It allows people to agree on a general goal and then to compromise on the particulars. A word like 'community' tends to become overused, underdefined, and ominously upbeat. (The cultural historian Raymond Williams once said: "It was when I suddenly realized that no one ever used 'community' in a hostile sense that I saw how dangerous it was."[11]) A crisis can strip away the useful but limited cover of ambiguity.

After the death of Scott Krueger, MIT went through a crisis of community. Many students and alumni/ae fiercely attacked MIT for destroying the fraternity system, depriving students of residential choices, or treating them like children. On the task force, although we had built personal relations of trust and respect during our months of working together, we still differed in our fundamental definitions of community. Because of our personal relationships, however, we did not just shout our positions at one another. We kept trying to work through the differences. It was a difficult period. In addition to our regular meetings, the co-chairs held a series of meetings with individual members to try to resolve our differences. Not far in the background, the Middlesex County District Attorney was conducting a grand jury investigation of Krueger's death. The Boston Licensing Board and its counterpart in Cambridge were taking measures that threatened the survival of fraternity houses and indeed of the entire fraternity system. Local political and community leaders kept asking how MIT could allow its first-year students to live in what they considered unsupervised houses. So did the local and national press.

By late spring of 1998, the task force had reached a consensus to include in its report a recommendation that all freshmen live in campus residence halls. It had not been arguments for or against "community" that had led the student members of the task force, with the Student Advisory Group behind them, to go along with the recommendation; it had been the sense that this change was inevitable. The way this came about reveals the deeply fractal pattern of community life: we live not in a hierarchy but in an interactive network of communities. While the MIT "community" was deeply divided over the virtues of the MIT housing system, the surrounding world— the "communities" of Boston and Cambridge—concluded that the Institute had been irresponsible in its supervision of student living "communities." Whether through licensing procedures, the criminal justice system, or insurance regulations, the demands of the larger community proved more potent than any debates within MIT. In particular, Boston and Cambridge authorities made clear their conviction that MIT freshmen had to live in a more supervised setting.

We on the task force felt that we could help MIT make sense of this outcome by putting it in the context of the need to build a larger community. The final report of the task force, signed by all members, was published at the beginning of September 1998. The "central finding" of the report showed that in the two years since the Cape Cod retreat we had risen far above the level of complaints such as those about 8.01:

Given the challenge of helping students develop the qualities of the educated individual, it is appropriate that the Task Force was asked to examine the interaction between student life and learning. *The Task Force's central finding is that the interaction among these elements of the student's experience is fundamental.* The combination of structured learning and unstructured or informal education is critical because it enables us to educate the whole student. . . . The central and distinguishing feature of an MIT education is that it incorporates the three elements of its educational triad—research, academics, and community—into an education that is greater than the sum of its parts.[12]

This "central finding" came directly from students, both those who were officially on the task force and those who worked with it in an advisory role. Their leadership, combined with that of the co-chairs, deflected the task force from the potential dead end of curricular reform to agreement on the value of human interaction as an essential part of education in the information age. In the pivotal fourth chapter of the report, the task force defined 'community' as "students, faculty members, staff, and alumni who have come together on campus for the common purpose of developing the qualities that define the educated individual."[13]

The emphasis throughout the report was not on building community through technology but on the opposite: what technology *cannot* do to create community. The task force gave priority to an MIT community based on geographical place, not on technological systems:

Each of us is an example to our peers and colleagues; through professional, recreational, and social interaction with one another we build a culture of discovery and learning. . . . Hence informal personal interaction can be considered the life of the "community": student activities, casual social get-togethers, cultural events, and daily encounters with friends and colleagues are a few general categories of such interactions.[14]

The task force acknowledged the common prediction that as "information technology introduces new methods for teaching and reduces the barrier of distance" it will challenge residence-based education.[15] The response of the task force was not to accept the death of the campus as inevitable but to define the added value of an MIT education in terms of personal, place-based experience. Other institutions may be better situated to offer distance learning; we can cooperate with them, but in doing so "we must not devalue human interaction":

MIT's contribution will be the way it brings together the best people with the best technology to produce excellence in education. We must focus on this goal, rather than on the technologies themselves.[16]

In this human-centered spirit, the report calls for "a committed, yet cautious, process of experimentation, evaluation, and dissemination" in distance learning and educational technology, with an emphasis on serving students on campus as well as any students off campus.[17] The report also notes that, although external initiatives in distance learning may seem attractive ways to expand the student base, they "may be inconsistent with MIT's principle of excellence and limited objectives."[18] In case anyone missed the point, the first recommendation in the chapter on "Strategy and Structure" was:

Maintain MIT's excellence by continuing to focus on education and research that take place on campus.[19]

In the final section of the report, the task force called for nothing less than a "cultural shift" at MIT:

*from* demanding separation of student life and learning
*to* demanding that they be inseparable,

*from* focusing on formal education
*to* emphasizing learning in both formal and informal settings,

*from* a community divided by place, field, and status
*to* a community unified by its commitment to learning,

and

*from* keeping research, academics, and community apart
*to* unifying the educational value each provides.[20]

The report concludes that adding "community" as an essential element of MIT education "is not just a necessity we must grudgingly accept, for it also opens bright prospects for the future. . . . Today's need for change presents the opportunity for another leap forward, and a chance to make MIT the same guiding light for higher education in the twenty-first century as it has been in the twentieth."[21]

## The Conflicts of Community

MIT people are smart, but they are generally not very intellectual in the sense of being interested in ideas. Most members of the task force were blissfully unaware of the intellectual and political complexities of the concept of "community," on which books have been written. Only a few of us were aware of the existence of "communitarianism" as a political movement intended to counteract corrosive individualism and neoliberal market orientation.[22] Still, many on the task force were uneasy about the word 'community' because its meaning was not self-evident, even to us. One self-described "triadophic" faculty member (not a member of the task force) reminded us that the word 'community' is incongruous with the other two legs of the "educational triad." How do you *do* community, he asked, in the same way that you do research and teaching? He suggested we say that MIT education focuses on "the vigorous pursuit of academics and research within the context of a diverse and dynamic community." No one really quarreled with his suggested phrasing, but its very precision kept it from having the punch of endorsing a new "leg" for MIT education.

One did not have to be an intellectual to note, especially during the first few months after Scott Krueger's death, that 'community' could refer to social groups of different scale—the city, the university, the living group—with no way of establishing priority among them. Generally, though, faculty members on the task force wanted to endorse a strengthening of the middle range, the institutional range. We also wanted to endorse a strengthening of communities based on physical presence rather than on virtual presence. Task force members were well aware of the distinction between a local, geographically based community and extended communities based on shared messages, interests, and values (often called "virtual communities" or cybercommunities" in the information age).[23] When the task force endorsed the value of education *on campus*, we meant, and we said this as clearly as we could, "Cambridge 02139."

This bias is curious, since universities are self-evidently founded on the principle of universality. Universities distrust local knowledge: scholarship is legitimized only by bringing it to the global store of knowledge, where its excellence can be evaluated by the best people in the field. Today's virtual communities can be seen simply as technically souped-up versions of the long-standing scholarly community—an international network of scholars, connected by canonical texts, shared language, and shared norms, that dates back at least to medieval universities. This international network has been refined significantly since the emergence, in the seventeenth century, of the scientific community, which has repeatedly developed innovative ways of knitting together the minds of its members through letters, papers, journals, peer review, visual display, rhetorical conventions, dispute resolution, ethical standards, and most recently the use of email and online publication. Today a faculty member's professional advancement depends on his or her reputation in these extended communities.[24] (In one tenure case reviewed by Academic Council, a dean praised the candidate as "a good citizen of the yeast community.")

But if knowledge is global, learning is local. As the task force forcefully argued, personal contact between teacher and student has been the core of higher education.[25] Excellence, universally defined, and learning, locally performed: a university tries to combine two types of community in one institution, and faculty members try to reconcile these dual allegiances in daily life. If no institution can endure without inner contradiction, then the contradiction between localism and globalism lies at the heart of the university, endowing it with both tension and energy.

In theory, such a fundamental conflict might tear a place apart. In practice, at MIT, the two sets of community allegiances coexist quite happily. There is an MIT of place-based communities, localized in living groups and departments and offices, measuring time in the rhythm of weeks and semesters. There is also an MIT of scholarly communities, extending to colleagues around the world, measuring time in a rhythm of grant writing, publication, and annual meetings. The task force's endorsement of local community was by no means a rejection of extended communities. We assumed their peaceful coexistence. From experience, we knew that there is no consistent correlation of local communities with warm fuzzy feelings and extended ones with cold impersonality. In a scholarly community many interactions take place across distance, but people try to meet regularly, both at formal meetings and in informal encounters. Even if they see one another infrequently, they can develop a rapport that is enhanced, not damaged, by systems-based communications. Conversely, even on a relatively small campus, human interactions may be irregular. At MIT both students and faculty members are better acquainted with their smaller neighborhood (the residence, the department, the lab) than with the rest of the academic village. As is true everywhere else, they are not always fond of their closest neighbors. (Just ask department heads and housemasters.)

The view of the task force was that regular local interactions had to be encouraged and enhanced in order for the Institute to continue to excel in its educational mission. The task force report proposes three new principles for MIT education, in addition to existing principles inherited from William Barton Rogers and the Lewis Committee: "an integrated educational triad of academics, research, and community," "intensity, curiosity, and excitement," and "the importance of diversity."[26] The three are tightly coupled. Experiences of "intensity, curiosity, and excitement" emerge primarily from face-to-face human encounters. The importance of diversity is also experienced in human encounters on a campus where the student body varies widely in ethnic, racial, religious, linguistic, and geographical identity. In fact, it could be argued that one of the best ways for an MIT student to become educated for global life is to live in Cambridge 02139 for four years.

This is overstating the case, to be sure. As already noted, the otherwise varied campus community is relatively homogeneous in intellectual orientation. Furthermore, there is no substitute for living, working, and studying abroad to get an intensive experience of the variety of the world, and MIT is more and more encouraging and enabling its students to do this. Nevertheless, the MIT campus offers an experience in lived diversity because students of widely varied background share a place where human differences have to be confronted and reconciled. Global networks of communication, on the contrary, often link people who share ideas and interests, so that their differences as full human beings are less evident and less problematic. Too often discussions of place-based vs. globe-spanning virtual communities assume that the local is relatively homogeneous, while the virtual is diverse. At a place like MIT, if anything the opposite is true, because there the local is so wonderfully globalized.

Most of us thrive at MIT on the mix of local and extended involvements. In this respect, university life may be a prototype of the social life of the future.[27] The challenge is to keep the mix a balanced one. In reasserting the values of "Cambridge 02139," the task force was implicitly issuing a warning that extended communities are undermining local ones. Why is this happening? Certainly the power of new technologies is a major factor. The dream of instant worldwide communications—a dream that dates back to the Enlightenment, and was a critical component of the dream of progress—inspired the construction of a web of communication and transportation systems that suck energy away from institutional centers. In the age of jet travel, faxes, email, and the World Wide Web, it is harder and harder to balance the demands of extended networks with those of more local networks.

But more than technology is involved in the weakening of place-based communities. The reward structures of faculty life are also changing, and in ways that work against local attachments. At several points the task force report cites "market forces" that are putting pressures on MIT: the rising costs of higher education, shifting sources of revenue, and pressures to expand the student base and the research base.[28] But the forces of capitalism (a word not used in the report) are even more far-reaching. The scientific community is not just an information network: it is part of the world economy. In the Cold War years, the funding structure for research depended primarily on national governments and economic nationalism. As national support for scientific research flattens, the growing sources of support are often corporate, and corporations are increasingly multi-national. Worldwide economic competitiveness encourages worldwide scientific collaboration. A truly global scientific community, and the existence of "extended labs," are being realized in practice. So is the emergence of a truly international job market for scientific and technical expertise.[29]

It was hard enough for MIT faculty members to balance the claims of the local residential village with those of the global scholarly village. When they add to these the claims of the global market, in the form of entrepreneurship and far-flung funding collaborations, they feel constantly torn by the multiple demands on loyalty, identity, and attention. When scholars chase funding in a global market, more and more of their time is spent piecing together complicated grant proposals or corporate partnerships, instead of relying on more predictable national sources. In a worldwide market where top scholars bring top dollars, they are more likely to entertain offers from other institutions. With the rise of venture capital, MIT faculty members and some staff are increasingly involved in entrepreneurial ventures that demand their time and attention. Some professors can earn much more outside the Institute than from their salaries. Their ventures are open ended in their demands, whereas traditional consulting arrangements were more clearly defined by the "one day a week" rule.

The task force, and MIT at large, were deeply ambivalent about global capitalism. MIT needs the resources of the global scientific marketplace to support its education and research. Those resources support a collective life characterized by "intensity, curiosity, excitement" and by mind- and soul-stretching human diversity. At the same time, the multiplicity of opportunities beyond the campus can be distracting. As distant connections and responsibilities multiply, they leave less time for the local milieu of innovation. Being busy is not the same as being productive, as many of us are all too aware. A life devoted to messages and information reduces the mental grazing room necessary for creativity.

At MIT today the yearning for local community is genuine, but it is constantly checked by the need to remain competitive in an academic and research marketplace, which demands a non-communitarian organization of time and space. While trying to build a shelter from capitalism, we are busy reinforcing its structures. The split runs right through MIT as a community, and right through the hearts and minds of many individuals there.

### The Crisis of the Lifeworld

A community is "a social entity created in space through time."[30] The major finding of the task force, stated positively, is that the MIT community is critical for human development at MIT. The major finding of the task force, stated negatively, is that the space and time necessary for community and therefore for human development are becoming ever scarcer at MIT. Scarcity is usually thought of in material terms, but at MIT today the worst scarcities are phenomenological.

The shortages of time and space at MIT are described most plainly in the fourth chapter of the task force report. That chapter enumerates both the strengths and the weaknesses of the present MIT community. It cites five strengths, all human ones: loyalty to residence, independence of community groups, diversity of existing groups, a strong athletics program, and the quality of the MIT staff. Then it lists eight weaknesses, all reducible, in the end, to lack of time and lack of space. The report lays out the problem of multiple and proliferating "faculty commitments" and "student commitments" in a world where time is finite. It lists community-wide functions that are short of space on a campus designed primarily for the needs of subgroups: housing, dining, social activities, athletics, performing arts.[31]

In making eight "community recommendations," the task force report tries to imagine how MIT might find more time and space for its community life. In many ways, finding space is easier, if expensive: it can be built. The report calls for an ambitious program of design and construction to provide more facilities for community activities.[32] But how do you provide time? "Of the many difficult design problems MIT faces, promoting student and faculty participation in community activities is probably the most difficult."[33] The task force proposed changes in the reward system to motivate such participation. For faculty members, for example, "community participation should be considered in the tenure, promotion, and performance review process as part of a faculty member's teaching record."[34] For students, community time can be encouraged by making orientation more of a shared experience, and by making the first-year program focus more on the larger community.

Of all the effects of technological change, the ways it alters time and space are among the most important and pervasive—but also among the most difficult to apprehend, understand,

and control. What changes with technological change? Not just the appearance of things, products, gadgets, structures, systems. The whole world changes. The human-built world is not a Newtonian universe of absolute time and space that gets more and more filled up with the things and systems we create. The technological world is relativistic, in the non-rigorous sense that time and space are altered by the presence of matter.[35] Innovations do not just pile up. Their presence distorts time and space, and the distortions are experienced daily by people living in a time-space field. The things we make also multiply spatial connections across distances. The things we make also multiply demands on time. The integrated effect of all these proliferations causes the feeling at MIT that we "suffer from innovation." A capitalist economy favors some types of innovations over others. In such an economy, profit is created by movement in time, and profit increases with compression in time and/or with extension in space. Since one important purpose of technological innovation under capitalism is to increase the value of movement in time, innovation exerts relentless pressure to increase speed and to intensify the use of time.

Capitalism has brought many crises of overproduction, but the crises have been economic, resulting from too few buyers for the goods being produced. The "things" of the information age—data, reports, communications, messages—serve social and psychological needs as much as they serve physical ones. When they are overproduced, their glut carries with it not just economic but also psychological and social effects. The proliferation of information is liberating: it brings rewards of "intensity, curiosity, excitement." But it also has the structural effect of underproducing time and space for collective life.

When there is (to paraphrase Latour's motto) "innovation without representation," the accumulation of innovations degrades the lifeworld. Market forces constantly press on time and space that individuals and institutions have reserved for other purposes. It is the twenty-first-century version of the enclosure movement, in which early modern capitalism created wealth by taking over property that belonged to others.[36]

The weakness of "community" is less a cause of the problem than a symptom of the underlying problem: the process whereby market-driven innovation structurally and relentlessly invades the time and space we try to reserve for personal and social life. It is a new tragedy of the commons, a social tragedy this time, when we discover that we are "not built for a world we have designed."[37] It is an environmental crisis, not in the usual sense of physical pollution of the non-human world, but in the sense of the degradation of the human world.

In his later works, the philosopher Edmund Husserl used the term 'lifeworld' to describe the human environment in something like this sense. Husserl was trying to describe, in a comprehensive way, the commonsense world of experience— the world we encounter in everyday things and everyday practices. Husserl wanted to contrast this supremely human world of semi-conscious, sense-based experience with the abstract world of scientific or philosophical reflection.[38] As so often happens, in making a distinction he was also making a value judgment. Husserl was asserting that the lifeworld, as "the unified ground of experience," takes precedence over "the abstract and derivative worlds of science that arise from it."[39]

A similar assertion could be made about technology. In the creation of a common world for human life, the unified ground of technological experience should take precedence over the abstract processes of technological change. All those abstract processes intersect and eventually come down to earth in everyday, concrete experience of a world that includes the full range of objects and practices, both human-built and given, as they are valued and used in both economic and non-economic ways.[40]

Many people at MIT—staff, students, faculty members—feel they are living in a lifeworld crisis, even if they do not have a label for it. The crisis is defined by a sense of crowding: things, communications, people, and information relentlessly accumulate in self-reinforcing structures of high demand and high visibility. Space is crowded. Time is crowded. Life is crowded. Attention is short. Patience is short. It is like living in an echo chamber, sometimes stimulating but sometimes deafening. The lifeworld is at once enclosed and full.

Daily experience, repeated and compelling, is the ground truth for how we think about the "abstract and derivative worlds of [technology] that arise from it." The sense of living in a crowded lifeworld becomes a more general sense of technological inevitability, in the experiential (if not logical) sense of feeling that "technology" is what determines the conditions of your life and also of the larger life of society. Technology is the future. Technology is history. Technology is the human destiny. How can we reclaim the lifeworld when its distortions result from forces that seem ghostly, even supernatural, and hauntingly elusive? How can we retool it?

## The Spatial Crisis

The task force report suggests that the shortage of space is an easier problem to solve than the shortage of time: all you need is lots of money. Or so it seemed: MIT's experience in trying to act on the recommendations of the task force suggests that it is becoming well nigh impossible to build a physical world with enough speed, flexibility, and capacity to keep up with technological change.

In 1995, when I became dean, MIT had 10 million square feet of space, compared to only 3.5 million square feet in 1960. If the rule of thumb is that a university needs a million more square feet every ten years just to keep up with the normal expansion of knowledge, then MIT had done more than its share of expansion. Nevertheless, by the mid 1990s everyone at the Institute was complaining about a space crunch. In a fractal pattern, from the top to the bottom of the Institute, from administrative suites to parking spaces, struggles to find, assign, maintain, and renovate space consumed huge amounts of time and energy. A staff member involved in these struggles told me, in exasperation: "It's just like a huge jigsaw puzzle. You can only pop out existing pieces and try to move them around."

At times it seemed that everyone was going around the campus with a tape measure, trying to squeeze the last few square feet out of the system. When the women faculty members in the School of Science began to suspect gender discrimination, one of the first things they did was get tape measures and compare their research space with that of their male colleagues. (They indeed had less.) The libraries did not have enough shelf space for books and journals, so more and more materials were sent off site for storage. Classrooms were tightly scheduled, especially the larger lecture halls. The Registrar's Office played the role of a maître d', only booking classrooms instead of tables. Every year, administrative heads (myself included) took measurements and drew floor plans to demonstrate how crowded and fragmented our office space was.

Student housing was in especially short supply. During the 1990s, the end of rent control in Cambridge and a brisk housing market in the Boston area brought a sudden halt to the long history of students' moving off campus to live in inexpensive shared apartments. The overflow backed up on campus in the form of a greatly increased demand for MIT housing. Some years, the number of on-campus "crowds" (four students in rooms intended for three, for example) was disturbingly high. As the number of graduate students steadily mounted, the percentage that MIT could house dropped far below the stated institutional target of 50 percent.

Above all, as the task force had emphasized, MIT was short of "community space." Arts groups lacked places to practice and to perform. The one good-sized auditorium on campus— which has engraved over its entrance the wonderful words "A Meeting Place for the MIT Community"—was booked years in advance. Each year student officers of the Association of Student Activities, which allocates square footage to various groups, made unannounced after-hours visits to rooms assigned to those groups to be sure that the space was in fact being used. Indoor athletic facilities were dingy. Playing fields with natural turf were booked every daylight hour and thus were hard to maintain. The artificial turf on other fields was worn and lumpy.

The space crunch was made worse by the fact that MIT, like many other universities, had accumulated a staggering backlog of deferred maintenance by the 1990s. Many classrooms had desks, blackboards, and lighting that dated from the time of the Lewis Report. Some of the shabbiest were located in the Main Group. Buildings put up in the 1950s and 1960s were beginning to need considerable retrofitting. The chemistry building, for example, needed millions of dollars' worth of work to come up to code in basic safety equipment.

Shortages of any kind do not typically bring out the best in people, and the space crunch was no exception. My predecessor as dean, Arthur Smith, perceptively warned me that money problems are much easier to solve than space problems. "Money can move around," he said. "With space, you have to move someone out." The information age has not removed the deep human instinct of territoriality. People do not like to move out. In the dean's office, staff would make jokes about setting up trailers in Killian Court or putting barges offshore in the Charles River. They were also enraged by the prospect of being downsized from an office to a cubicle, especially when their work involved confidential conversations. In the dorms, housemasters spent hours with roommates who could not find ways of sharing space successfully. In the research area, department heads and deans spent hours trying to negotiate "space wars" among faculty members. Space was the limiting factor in any request to expand programs in a new scholarly direction.

In sum, MIT struggles with the production of space.[41] Its ability to produce technological and scientific ideas for the great wide world outruns its ability to provide itself with a place in the world to continue to generate ideas. There are many reasons for the space crunch that become so acute in the 1990s. Many of them have to do with forces over which MIT had little or no control, such as changes in overhead policies and the end of rent control. With that said, another source of the spatial crisis has been the deterritorializing effect of the information age. Production of virtual space, it is often assumed, should relieve pressure on the production of physical space. If you can work and learn in cyberspace, for example, then it would seem you do not need so many offices or classrooms. You can work and study from a distance. This is the technological determinism that drives many of the predictions of the death of the residential campus.

But MIT experience demonstrates, if anything, the opposite. From a university planning perspective, there is little that is virtual about the information age. Increased production of virtual space does not reduce the demands for space in Cambridge 02139; rather, it multiplies the needs to build or renovate physical space, while making that space considerably more expensive. At the same time, attention to and investment in virtual space reduce attention to and investment in physical space. This diversion was abundantly evident in Reengineering. One staff member with significant responsibilities for space planning vented to me about a meeting he had attended where Reengineering consultants had held forth on SAP. He had asked them to explain how they were going to handle changes in physical space required by the new software system. They had not even thought about such an association. One consultant had responded: "Oh, we don't bother with that." The staff member glowered in recollection: "I was ready to lean over the table and strangle him! He makes three times as much as I do, and he can't answer that basic question!" In one memorable reengineering discussion about SAP, a participant stopped in midstream and commented, to no one in particular: "Do you notice how our discussions of virtual space always end up in physical space?"

At MIT we are still trying to figure out how to combine what the sociologist Manuel Castells calls "the space of flows" with "the space of place." We are beginning to do better, but over and over again the work of integrating virtual and physical space ends up being expensive in dollars and complicated in human interactions. For example, as dean I was involved in trying to integrate virtual and physical space in classrooms. When the administration made some funds available for classroom renovation, I organized and chaired a Classroom

Advisory Committee, more or less equally divided between staff and faculty, to oversee the design work. According to several surveys, most faculty members wanted only the most basic improvements: soundproofing, better heating, better light control, fresh paint, more functional furniture. When asked to list their wishes for audio and visual equipment, they requested better chalk and overhead projectors. In other words, most faculty members wanted what realtors would call "paint and paper." We on the Classroom Advisory Committee were not zealots for information technology. However, we felt that it would be wise, looking to the future, to consider how investments in virtual space might be made at the same time as the "paint and paper." For starters, we considered what kinds of audio and visual equipment should go into each type of classroom. In the not-so-old days, "AV equipment" meant overhead and slide projectors and screens. Now it includes computer projection, television sets, videocassette players, and DVD players. We decided that some classrooms could be equipped with a very limited range of AV equipment, but not many. Even in smaller rooms, a professor might want to use fairly sophisticated equipment.

So we started outfitting the rooms with electronic aids, which led to a cascading series of other costly decisions. If we were laying a new floor, we might as well go ahead and put in trenches below it so we could provide ducts for wires and outlets for laptops. If we were going to put in new light fixtures, they should be adaptable for several levels of illumination, depending on what technology was being used. Blinds had to be seriously rather than casually light-blocking. Switches for lights and AV equipment got complicated: we went with electronic rather than mechanical switches because they were more adaptable. For student seating we decided on tables rather than chairs with swinging arms, because arms do not readily accommodate laptop computers. We also discovered that we would have to contribute funds to provide better information-technology connections for entire buildings, since building connections were paid for by piggybacking their costs onto the price of any such internal renovation project.

In sum, we soon became carried away by another form of technological drift, this one set in concrete. In the first year, we invested nearly $2 million in the renovation of only three classrooms. As we renovated more classrooms, we would have more economies of scale, but it was still far more expensive to renovate than in the pre-information-technology era. The price kept rising in part because it was so difficult to predict what technologies faculty members would actually be using. The cutting edge tends to be the bleeding edge, financially speaking. It is easy to overinvest when all possibilities seem open and when choices are unclear.

Once again, too, technological drift had the effect of redefining work. For faculty members, the classroom is the shop floor: it is where they do their teaching work. In going beyond their expressed preference for "paint and paper" renovations only, in providing them with a new kind of "shop," the Classroom Advisory Committee was in its own way pressing change on those resistant to change. We were encouraging faculty members to try out new technology—to change, as the saying goes, from teaching as a "sage on the stage" to teaching as a "guide on the side."

And, once again, the effort to redefine other people's work through technology met with resistance. In the case of faculty members, it was passive resistance. Not by accident, the very first round of classroom renovations (carried out before the formation of the Classroom Advisory Committee) took place in rooms used by the math department. It had a reputation for a conservative teaching style, and the hope was that the presence of new gadgets would encourage innovation in teaching methods. Many members of the math faculty, however, just ignored the new bells and whistles. Three years after the classrooms were renovated, the chairman of the MIT Corporation, Alex d'Arbeloff, gave $10 million to the Institute to encourage innovation in teaching. The math department submitted a proposal that would allow a group of faculty members to commit significant time to innovative projects such as chat rooms, help lines, and self-teaching modules. The math faculty wanted to enhance self-paced student learning outside the classroom, in order to reduce dependence on the classroom as a primary learning site. They were not resistant to change after all; rather, they valued time over space as an incentive to change. When they did, the expensive classroom renovations were still underutilized, because the faculty had decided that classrooms were not necessarily the most important site of learning in mathematics.

Somewhat more surprising was the resistance of some undergraduates as they began to be taught by technology-intensive methods. The conventional assumption is that a computer-literate generation prefers computer-intensive teaching methods. The conclusions of the task force, a student-driven process, challenge this assumption: its report concludes that MIT students want human interaction more than computer action. In a "Letter to Faculty" dated April 11, 2001, the chair of the MIT faculty and the chairs of the two most important committees dealing with undergraduate education cautioned their colleagues:

. . . we are finding that some new educational initiatives conflict with long-standing MIT practice. We are especially concerned that online delivery of subject material and asynchronous discussion is being substituted for face-to-fact contact between faculty and students and among students in a cohort. . . . The recent *Task Force on Student Life and Learning* report emphasized the central importance of personal contact throughout a student's years at MIT. We endorse this finding and intend to be particularly careful when proposed subject changes would reduce actual contact time between faculty and students.[42]

Students echo this plea for human contact, even if it takes place in a lecture hall. In the fall of 2000, Web-based lectures were substituted for the traditional lecture-and-recitation format in 6.001 (Structure and Interpretation of Computer Programs), a popular initial course in computer science. Faculty members and administrators often assume that students find lectures boring, passive, alienating, and ineffective. The new version of 6.001, however, prompted an editorial in *The Tech* (MIT's student newspaper) complaining about the "impersonality of the Web." The editorial writers complained that Web-based instruction does not permit faculty members and students to make adjustments and observations in real time, based on sometimes subtle clues. The writers added:

Perhaps a more troubling effect of moving lectures to the web is that the class undercuts what many MIT students, faculty, and administrators have sought to build for so long—a stable, vibrant community. Instead of going to lecture, asking questions, and discussing problems with fellow students, those enrolled in 6.001 may now sit at home, alone in their rooms, and receive lectures with zero personal communication by staring into a computer screen. A critical component of MIT community and camaraderie is the shared experience of attending lecture with one's peers. 6.001's system guts this very important aspect of MIT life.[43]

The editorial went on to recommend that Web-based tools be used as supplements for traditional lectures rather than as replacements for them. Since MIT undergraduates are arguably one of the most technologically literate groups of young people in the world, their "resistance" drives home the point that in most cases the cause for such resistance is not fundamentally technological.

Nevertheless, everyone understands that the technology of teaching and learning is changing, and often for the better. As this change happens, it puts into motion a series of requirements for new kinds of physical space. As the math faculty understood, classroom habits may change less quickly than homework habits. In the age of lectures, recitations, problem sets, and textbooks, students often did their homework in the library. They still do—but with the increase in student population, and the crowding of libraries from overflowing shelves and banks of computer terminals, students complain that they cannot find study space there. This is especially true as homework assignments become more often team-based. Students can do online work at a computer cluster, but the clusters are often crowded and are not conducive to teamwork (the computers are placed along long tables, back to back). There are team study rooms in some of the dorms, but they generally are not equipped with online connections. There are few project-based team study rooms where students can work on something in physical space, leave it overnight, and come back to work on it the next day or week.

Students need access to the virtual world, and they need access to other students. Too often they have to choose one or the other. In the meantime, collective spaces for doing homework being less than ideal, many students end up working in their rooms. At MIT, this was encouraged, not with particular forethought, through an institutional decision to invest in ResNet, a project to bring the Unix-based campus network connection into individual rooms. As a result, students' rooms now accommodate computers and printers in addition to sound systems and cooking appliances. Students spend more time in their rooms because the electronics are there.

Once again, the space crunch does not bring out the best in human nature. One faculty housemaster commented that a typical freshman dorm room "has become not much different from living in a submarine, from the perspective of human dynamics." He is convinced that the "submarine" environment has resulted in more frequent and more serious roommate problems, as students spend more time together and compete over both physical and virtual space: "There is also the problem that (from what the students tell me) the protocol used to split the ResNet connection in the dormitory room that provides 'one port per bed' will not allow the simultaneous use of the network if a portable computer is attached. Thus, if one of the roommates has a portable, that student monopolizes the ability to work in the room."

These downstream effects of new educational technology are solvable. MIT's new undergraduate residence will have four data ports per bed and a network computer cluster designed to foster conversation and interaction. But the cost of retrofitting old dorms is out of reach, in view of the need to attend to basic safety and fire issues first. Ten years ago, we defined classroom renovation as a discrete project with its own budget. Now we recognize that the whole campus is involved in the renovation of learning space, and that the cost goes far beyond that of installing computers and networks in classrooms.

MIT is now addressing its spatial crisis as it enters a phase of physical renovation unmatched since the 1950s. Long-deferred maintenance is being attended to. Classroom renovations continue. New research buildings are going up or are about to go up: the Stata Center for computer and information science, a building for neuroscience, a building for management science, an extension of the Media Lab. Other projects respond to the task force's call for community space: a new sports and fitness center, performance space, and an exciting "student street" as part of the Stata Center. In addition to the construction of the new undergraduate residence, Simmons Hall, a capacious graduate residence is being built.

For many years, the campus is going to be a vast building site. Each week, *Tech Talk*, the Institute's official newspaper, carries a special column telling people what construction is happening and which parts of the campus to avoid. Each day brings a new pit, a new obstacle, a surprise. In one grassy quadrangle, a grove of trees is leveled. In another grassy quadrangle, trailers pop up like mushrooms. Parking lots are shut down to serve as staging sites. Some roads narrow, others change course. Cranes loom, drills thump.

This construction boom is in some ways the logical successor to Reengineering, which put into motion all sorts of changes that require new configurations of physical space. Reengineering and rebuilding are both projects that have been necessary, expensive, constructive, and disruptive. The amount of money being invested in physical space, however, dwarfs the investment in Reengineering. The price of computer chips may fall exponentially, but the price of building a world for chips seems to rise exponentially. Some problems can be solved, at least partly, by throwing money at them. Providing more space is one of them, but the scale of the spending is almost beyond belief. The total price tag for the projects MIT has in the ground and in design is at least a billion dollars.

Even spending on this scale leaves a long list of unmet needs for more and better physical space—for more student housing, for proximate faculty housing, for libraries, for performing arts rehearsal and performance space, for learning space.

Furthermore, in view of the pace of change in information technology, it is entirely predictable that the buildings being constructed or renovated will need renewal soon. Physical space is caught in an apparently unwinnable race to keep up with the evolution of virtual space. The conquest of space can be accomplished only through the production of space—but then the fixity of the space that has been produced becomes an obstacle to the next round in the conquest of space.

The rounds of restructuring are becoming ever shorter in duration. It was only a few years between the time MIT was busy opening campus network clusters and the time it decided to design wireless areas so clusters would not be needed. It was only about two years between the time we renovated a classroom with high-tech "bells and whistles" and the time we decided to undertake curricular reforms that downplay the role of the classroom. The mantra of continuous improvement does not take into account the costs, financial and otherwise, of providing built space. As the geographer David Harvey notes, "the production, restructuring, and growth of spatial organization is a highly problematic and very expensive affair, held back by vast investments in physical infrastructures that cannot be moved, and social infrastructures that are always slow to change."[44]

As the cycle speeds up, the production of physical space becomes more and more of a drag on the production of virtual space. In a world dedicated to technological change, this is an inherent and increasingly limiting contradiction. Human connections through messages evolve faster than physically based connections. The virtual world cannot exist without the built world, but the built world cannot keep up with the virtual one. "Technological change" may find its choke point in the production of space. Or maybe it will be in the production of time.

## Time Scarcity

In public and in private, Bob Silbey and John Hansman, co-chairs of the task force, repeatedly stated the conclusion of the task force that the scarcest resource at MIT is not money, not even space, but time. "We used to be an exciting place," one member of the task force lamented, "Now we're just busy." The same thought was often expressed in Reengineering circles, whether in speech or on stickies. Everyone at MIT talks about the time crunch. When a group of faculty members pressed President Vest to put money back in the budget to reopen the Faculty Club,[45] he wondered out loud, "If we find the money, will anyone have the time to use it?"

Thomas Hughes has often said that we express our values, both good and bad, in our technology and architecture. The same is true of our calendars, the scaffolding of our days. Each person at MIT, from top to bottom, is engaged in a constant temporal cost-benefit analysis in which complicated and ultimately incommensurate values are weighed and sorted to reach a precarious personal compromise. Faculty members weigh civic duties against exercise, family time against work time, committee service against research work, local community against extended community. For students, the analysis weighs time investment in a technical class against investment in a humanities class, or both against some form of student-centered recreation. The "felicific calculus" has invaded everyday life far beyond decisions we think of as market-based ones.

Like people elsewhere who lead similar lives, MIT people depend on clever ways to appropriate time—snatching moments to chat in the hallways or rest rooms, reading email while they chat on the phone, reading hard-copy mail in committee meetings, always multi-tasking and shoehorning. They use an array of technological fixes, ranging from the Palm Pilot to the Web-based "student stress minimizer" that links a student's registration with class syllabi in an attempt to smooth out the workload over the semester. The students are the real masters of time management. They combine and juggle requirements, figure out how to sequence work on assignments, and sign up for a large number of classes and then drop them at the last possible moment if the combined workload becomes too heavy.

In most cases, students do all this time management in a candid and straightforward way. Unfortunately, in other cases, time scarcity at MIT—like any other shortage—causes dissension and resentment. For example, the student stress minimizer depends on the cooperation of faculty members, who have to be willing to shift homework assignments, paper deadlines, or quiz dates forward or backward by a few days to smooth the student workload. Some faculty members would be willing to do this, but by no means all. As the report of the Student Advisory Group of the task force explained, while students often feel that faculty members should spend more time in personal interactions with students, faculty members often feel that students spend too much time on their own time-wasting activities (each faculty member has his or her favorite culprits: parties, computer games, athletics, . . . ).

Furthermore, faculty members blame one another for taking up a disproportionate amount of the student's time budget. Nearly all curricular discussions boil down to competition for the student's time. Because faculty members are not able to agree on priorities, they all throw their requirements onto the heap, and the result is the curricular logjam. Scientists are sure the major requirements of the engineering departments are the main source of curricular overload. Engineers are sure the heavy General Institute Requirements (including the new biology requirement) are the problem. Humanists blame both science and engineering for heaping so many problem sets on the students that they doze through discussions and dash off papers at the last minute. The MIT curriculum is an educational commons that has been severely overgrazed, the result being exhaustion not of land but of students.

This situation worries everyone, because it leads to a sort of Gresham's Law that bad time drives out good. MIT's lifeblood of creative work, for faculty members and students alike, depends on having two different types of high-quality time: time for intellectual grazing, when random and apparently disconnected ideas are brought together in new ways, and time for prolonged and intensive work on ideas and projects. Time that is cut up by multiple demands, or cut across by multi-tasking and incomplete attention, is generally less productive. Weed time seems to keep spreading, driving out the better varieties.

On the task force we understood that we were describing the problem, not really solving it. Our report contends that increased time commitment to community is essential for MIT's educational leadership, but provides no convincing mechanism for providing that time. We did ask the Institute to "recognize" student and faculty participation in community activities in the form of notations on student transcripts, or to have community participation considered a part of a faculty member's teaching record in tenure, promotion, and performance reviews.[46] These are weak recommendations compared to the dominant reward structure, which is based on teaching as judged by the department and on research performance as judged by the extended scholarly community. After the task force's report, as before, community time remains marginal pro bono work, undertaken by a limited pool of dedicated faculty "good citizens" attracted by a hard-to-define set of rewards, including the hope of "making a difference." That is why the task force report ends with a call for a deep "cultural shift." If you feel it is well-nigh hopeless to change socioeconomic structures, you can always call for cultural revolution.

Two years to write a report! Governance, whether of MIT or of our society more generally, depends on investing time in decision making. Here it is difficult to achieve significant gains in productivity. Gathering relevant information and reflecting on it are activities that cannot be greatly compressed. Accordingly, institutions and individuals often do an informal cost-benefit analysis and figure that the consequences of making a less-informed decision will cost less than spending much more time to arrive at a well-informed one. This is the paradox of "the rationale of growing irrationality," by which quality of decisions declines because the processes of argumentation, negotiation, reflection, discussion, compromise, adjustment, and response are so woefully inefficient.[47]

But there is no other way to compose a common world. In a reflexive world, time is as reflexive as anything else. How we invest in time shapes and reinforces our future investment. When time and space are in short supply, community life suffers and shortages of time and space become even greater. There is an old saying that it takes money to make money. Similarly, it takes power to produce power, and it takes civic time to make civic time.

Provost Bob Brown once commented that "community has a cost: it's time." The only way to have community time taken seriously is to pay the cost. But how does an institution do this, especially when the people there are so tied up in networks of achievement that extend beyond the institution and over which it has no control? At least in theory, the space crunch can be relieved by spending huge amounts of money. There is no obvious financial solution to the time crunch, however.

Time scarcity is psychologically much more complex than space scarcity. Space is perceived as material and external. It is someone else's problem: if it is not provided, you can get angry and frustrated, but you do not typically blame yourself. Time is different. Time is you. When you are short of time, you scrutinize your priorities, then run through a private cost-benefit analysis, then make some accommodation between your own desires and your obligations. Time management is so stressful because this internal process involves such an array of subconscious desires and guilts.

The most serious obstacle to reducing the time crunch at MIT is the conviction of so many people there that it is inevitable because its sources are internal, not external. We have constructed a silicon cage of internalized discipline. In the task force's many discussions about the time shortage, over and over again faculty members (and students too) said that they drove themselves relentlessly, and that even if they were somehow given more time they would continue to drive themselves. In our truly honest moments, some of us admitted that our lack of attention to "community" was due less to a lack of time than to other priorities. We really like to teach, learn, and do research. Without a "cultural shift," the internalized drive toward individual priorities was not likely to change, and no one could see where that shift might come from. Even if MIT were to figure out how to pay the cost of buying out time for "community," would the offer be accepted?

MIT staff have the same tendency to internalize the problem of time scarcity. At one all-day Reengineering retreat, the facilitators concluded by asking people what one change in the reengineering process would be most helpful. As usual, stickies were passed around for people to write on and post on the walls. Most of them ended up bearing single word: TIME. Earlier in the day, we had engaged in a discussion about time scarcity, which had led the participants to list possible remedies: stress-reduction techniques, cell phones, massages, and, of course, more lists. I commented that we had identified a collective problem, which might conceivably have collective solutions. So far all the suggestions were individualistic in nature: should we not try to think of common remedies? There was silence, some nodding of heads, and then the discussion reverted to individual responses for "coping with stress."

Most people at MIT do not think of these feelings and experiences as political issues. Staff, faculty members, and students continue to internalize and individualize the problem of time scarcity. In fact they often resist efforts to address it by a sort of defiant addiction argument: "That's the way we are at MIT. We love what we do and whatever you do, you can't stop us from working hard." When Paul Gray was installed as MIT president in 1980, in his inaugural address he famously called upon the MIT community to take steps to reduce its deleterious "pace and pressure." When he retired ten years later, he just as famously remarked: "I didn't lay a glove on it." These two incidents are regularly recounted in discussions of the time crunch to demonstrate its intractability. This is MIT. We will always be stressed. Just try to stop us.

I respond, quoting Karl Marx: "Economy of time, to this all economy ultimately reduces itself."[48] It takes a combination of institutional parochialism and macho posturing to deny a connection between the time crunch at MIT and larger trends of time compression everywhere apparent in the economy. We have always prided ourselves on long hours and hard work, but these are no longer unique. Everywhere today you hear the same laments about lack of time for enjoying life, family, and community.

One major economic driver of the time crunch is the much larger trend toward defining work as an accumulation of roles rather than as a series of tasks. Tasks can be completed; open-ended role playing is never done. This redefinition of work is evident in the new world of Reengineering, where workers are encouraged to see themselves as free agents, moving from one team to the next. It is evident in engineering education, where entrepreneurship rather than job holding is the new ideal. And it is evident in the continuous expansion of the faculty role at MIT, as each professor tries to manage the integrated lifeworld effects resulting from multiple demands on time: research, teaching, private life, and now the MIT community. Occasionally the task force talked about trying to get a grip on this sprawl by writing a faculty job description, but we would have been swimming against the current in a world where employment in information-based work is defined less and less by "jobs" and more and more by ability to adapt, expand, shift, retool.

The protean and open-ended nature of the faculty role is, for better or for worse, a portent of the way things are going generally in the world economy. Such roles offer flexibility and variety; however, as open-ended hybrids, they place ceaseless demands on individual who are unable to distinguish the internal world from the external one and who, with mixed self-praise and self-blame, regard themselves as the source of their own busyness. MIT faculty members offer love of their work as a reason for putting up with the degradation of the common world. They are, in the words of the economist Nancy Folbre, prisoners of love.[49]

Where have we heard this before? The other job that people do because they love it, they say, sort of, is motherhood—a priceless source of joy that, when you cost it out, carries a huge price tag for women in lost wages and opportunities.[50] It has taken the women's movement decades to develop the conceptual tools to begin to address the "prisoner of love" argument. Self-exploitation is still exploitation, just as in Reengineering self-Taylorizing is still Taylorizing. In developing a new Politics in Latour's sense of the term (i.e., the progressive composition of a common world), what has been learned in the women's movement is crucial to success.

The appeal to faculty members to devote more time to "community" when no tangible reward is offered sounds eerily familiar to women. A faculty survey done by the MIT Planning Office in 1997, which became an important part of the "information gathering" of the task force, shows that female faculty members at MIT are more likely than their male colleagues to be asked to set aside time for the MIT community, and more likely to agree to do so. They also feel significantly more stressed in their professional lives.[51] It is not clear from these results if women feel this way because of greater attention to family life or because of greater susceptibility to guilty feelings. It does not matter. If MIT or any other institution relies primarily on "good citizenship" to motivate people to set aside "community time," women will respond, and suffer, disproportionately.

The dilemma is that "building community," to use that favorite phrase of the task force, makes demands on the lifeworld. Each human link one tries to make, each connection, each message, each effort to reach out and touch someone, happens both in space and in time. The framework of the lifeworld will continue to degrade until it is recognized that the provision of common time and space is part of Politics. The state of lifeworld consciousness today is similar to that of feminist consciousness in the 1950s and the early 1960s, when there was diffuse angst, anger that had nowhere to go, and vague awareness of a "problem that has no name." In confronting the crisis of the lifeworld, we are just beginning to understand, again, that the personal is the political.

# 5

# Men and Women in a Technological World

## How Gender Discrimination Works

This book's subtitle describes it as a historian's confrontation with technological change, but the historian is also a woman, and this matters. I could go back to the beginning and run it all over again to emphasize how much it matters, beginning with the girl in the sitting room listening to the men in her family (and only the men) talk about engineering. I was intrigued by their stories as a door into real life, but it never occurred to me, as it almost surely would have occurred to a boy in that situation, that I myself might walk through that door. Once in a while my male relatives would mention a woman in the world of engineering, but such a creature was rare and to me never seemed quite normal.

For all the time my male relatives spent chewing over the fate of the engineering profession, not once do I recall their discussing or even noticing the absence of women in it. This is how discrimination works in a relatively open and free society. We are unaware of it because it is not expressed in words or gestures of prejudice. It is just part of the ongoing order of things. Professor Nancy Hopkins has said that for her one great revelation has been to see how gender discrimination is practiced by very good and very smart people. She also observes that racial and gender discrimination seem similar in the way they can go unnoticed by those practicing them. Pointing out discrimination is, in most cases, not an accusation of personal misconduct but a description of deeply rooted psychological and social structures. Good and smart people grow up and live with most of the same psychological and social structures as everyone else.

No one ever told me that women were inferior, or that women were unable to do science and engineering. I was taken seriously as a person and encouraged to get a good education. I was also explicitly cautioned that women who had serious careers—who became college deans or presidents, to name two examples mentioned to me—usually remained unmarried, and usually remained childless if they did marry. Gender discrimination was built into the structure of expectations: men not only could but should combine a professional life and a family life, while women had to choose between them. I could not imagine being so ambitious about any job as to give up the chance for a family. I could have a normal life or a freakish one. This did not feel like a real choice; it felt like a message that I should not plan to take my work as seriously as men did.

Engineering was part of the world of serious work. Not surprisingly, the result was that the engineering profession was overwhelmingly populated by males, and engineering culture was defined by a male perspective. Language gives it away, always. When those of us on the Task Force on Student Life and Learning were reading and rereading the Lewis Report, we joked about just reissuing that fifty-year-old report with our names on it, because we could not possibly do better and our readers would be impressed by our brilliance. But we all were struck by one jarring note in the report, the one thing that gave away its date: the universal use of the pronoun 'he' in references to engineers and to MIT students. I have a similar awareness of cultural change when I recall that when I was doing research for my grandfather's talk on "Man's Use of Energy" I never even noticed the title.

But human life is not determined by social structures, no matter how powerful. Gender discrimination is not simple because socialization is not simple. Socialization also involves unpredictabilities of historical and personal circumstances, which can intersect at odd angles with the structures. Grandpa Lewis loved history and literature just as much as I did, mostly because of his love and respect for his cousin Mary, the English teacher. He passed on their shared passions to me. In fact, he told me more stories about the development of English democracy than he did about engineering. When I began teaching in a Florida high school, he was full of advice about books to assign my students. He suggested *Treasure Island*. When he was about twelve, he told me, one rainy afternoon he discovered a copy of *Treasure Island* in the attic of the Spring Garden farmhouse and stayed there reading for hours, spellbound. He often quoted lines from Shakespeare. Whenever he took one of his long train trips, my mother told me, he took along one of Shakespeare's plays from a set given to him by Cousin Mary.

The Lewis Report eloquently states the need for a "broader educational mission," but it never fully explains the deepest sources of that conviction, at least as far as my grandfather was concerned. True, World War II had demonstrated the need for engineers and scientists to understand the larger social context in which they operate. For Grandpa, however, the reason for a broad education lay in the human soul, not in social need. His ultimate concern was not solving problems, but understanding the mystery of human life. He loved history and literature because they told stories that provided insight into the complexities of human personality. He became interested in early childhood psychology, and more than once he expressed regret that he did not have a better understanding of it when he was raising his own children. Most of all, he looked to religious faith for guidance. Throughout his life he tried reconcile the teachings of Jesus with the principles of scientific inquiry, for he believed that Jesus had provided unique insight into the human mystery. For years he taught Sunday School in his Congregational church, and in his waning years he spent hours with an MIT chaplain talking about the reconciliation of scientific and religious faith.

Because Grandpa Lewis was so seriously engaged with history, literature, psychology, and religion, I never thought of them as specifically feminine, nor did I think they were incompatible with the engineering profession. The liberating force of his personality more than compensated for any social constraints of gender discrimination—constraints that are far more limiting for women who do not enjoy the personal advantages that I did.

In recent years, MIT has become increasingly conscious of its own structures of gender discrimination and has begun taking steps to change them. Until then, MIT's most important step had been its first and its least publicized: dramatically increasing the number of women undergraduate students. This shift was initiated, around 1970, by a small but influential group of faculty members and administrators who felt that they were doing something important for the future of science, engineering, and MIT. They understood that tokenism was bound to fail. They estimated that women would have to make up at least 20 percent of the undergraduates in order to flourish. That critical mass was achieved years ago. In the latest freshman class, close to half the students are women. The numbers are somewhat less impressive for engineering majors: one-third overall are women. But even here there are some interesting changes. I am particularly happy that the pool of undergraduate majors in chemical engineering is now 62 percent female.

Also around 1970, MIT began a concerted effort to increase the racial and ethnic diversity of its undergraduate student body. Now white males constitute less than half of MIT undergraduates, and one-third of the undergraduates come from families where a language other than English is spoken at home. When MIT people discuss the Institute's postwar history, they almost invariably point to the increase in diversity among undergraduate students as the most significant change. Once again this is a reminder of the vast difference between the banalities of change management, commonly defined as new software, and genuine historical change, defined as new ways of living together.

Today, MIT people usually describe the effects of increasing the proportion of women undergraduates in positive terms, and often in glowingly positive terms. It was not always so. In the 1970s some faculty members and others worried that admitting so many women, and in fact any effort to "broaden" the student body, would lead to a decline in academic standards. That worry has disappeared, since women's academic performance is just as good as that of men or even slightly better. Nevertheless, one still sometimes hears the complaint that women students "just" get good grades, without necessarily being creative and entrepreneurial. This is a classic example of changing the rules as the game progresses. Back in the 1950s and the 1960s, MIT tied undergraduate scholarship aid to grade point average (GPA), and fraternities proudly advertised their GPAs to show that they had met the critical standard of success at MIT. Even today students and faculty members put enormous emphasis on the GPA, which remains a decisive determinant for entrance into graduate school. The complaint that women are insufficiently "creative" and "entrepreneurial" may be psychologically useful to those who feel uncomfortable about the presence and the success of women, but there is no measurable evidence of the supposed deficiency.

The disturbing fact remains that even in the academic track, where being a good student is self-evidently a good thing, the number of women in science and engineering drops off steadily after the undergraduate years. This is dramatically evident in the MIT "pipeline" data. About 23 percent of graduate students in engineering are women, a sharp drop from the 35 percent of undergraduate engineering majors who are women. MIT's engineering faculty is now about 10 percent women. This is up from 4.8 percent in 1990, but it is still a dismal figure, especially because these 35 women (out of 348 tenured and non-tenured members of the School of Engineering faculty) are not evenly distributed. The numbers are particularly low in some of the largest departments.[1]

Moreover, too many women who have finally achieved senior faculty status at MIT find that their careers are contracting rather than expanding. When women faculty members in the School of Science were surveyed, every senior woman told a story of progressive frustration and marginalization. The details of the stories differed, but the narrative arc was always the same. One other piece of compelling evidence that indicates a problem is the absence of women at the very top of the institutional ladder. When I was dean, I was the only female academic on Academic Council. (Now there are two)

Clearly, gender discrimination has by no means disappeared at MIT or in the larger world. Most women are no longer explicitly told that they have to make a choice between family and work (or, more precisely, between a balanced human life and professional success as defined in today's marketplace), but they get the message anyway. The various women's committees now operating at MIT have spent a lot of time negotiating access to salary data, to ensure that men and women are being paid equally. The pay disparities they have found, however, are minor compared to this disparity: in the School of Engineering, approximately half of the women faculty members have children, while four out of five of the men do.[2]

Women do not drop out of engineering, as they go on in life, because they are worried about being paid less or getting less in benefits. They drop out because the more they look at the world they are headed for, the less they want to live there. When MIT's engineering departments offer faculty jobs to men, only 14 percent of the men decline the offer. When they make such offers to women, 40 percent decline. It takes stories to understand these data, and the stories tend to be about women not feeling at home in MIT's culture, about not feeling wanted there. These issues are hard to articulate, but the cumulative effect is striking and quantifiable. In a 1995 survey of the MIT faculty, three items stand out as showing pronounced disparities between the experiences of female and male faculty members. First, 34 percent of the women were single, versus 11.4 percent of the men. Second, about 35 percent of the women had considered leaving academic life, versus about 18 percent of the men. Third, *all* the women reported extreme or moderate stress (versus 88 percent of the men), and a far higher percentage of the women reported *extreme* stress (67.4 percent, versus 31.1 percent for men).[3]

Gender discrimination today is not a problem of women being excluded from engineering, by and large. It is a problem of women excluding themselves because, to use a phrase I hear from women all the time, "Who needs it?" When I talk with undergraduate and graduate women students about combining family and professional life, they tend to say simply "I want a life."

The feminism that matters today, for this subset of women, is feminism as humanism: the right to a life, as it were, as many women want to define a fully human life, rather than being forced to accept a male definition. Increasingly, women are questioning whether there might not be alternative models of work to make this possible. More and more, they are finding that men are allies in this quest. While I was dean, male students did not often come to me for family-and-work conversations, but they did often tell me about such conversations with their partners as they struggled to build a common life. Moreover, in the past few years I have had many conversations with male colleagues about raising children and dealing with aging parents—something that hardly ever happened in the 1980s, when I came to MIT. The crisis of the lifeworld, which creates scarcities of space and time, has affected women disproportionately but increasingly affects men too.

In 2000–01, a committee in the School of Engineering prepared a report on women faculty members similar to the one written five years earlier in the School of Science. The report presents statistics and analyses of salaries, hiring patterns, promotion patterns, committee service, outside professional activities, and other measures of institutional rewards and incentives. It forcefully demonstrates how male-dominated engineering remains. As with the report in the School of Science, the data are only part of the evidence: understanding the problem also requires weighing the cumulative stories of women's invisibility even at the level of professor—such significant details as, for example, never being asked to serve on Ph.D. committees for students in their specific area of expertise.

But the conviction that engineers solve problems—even problems as difficult as this—is just as much a part of the culture of little-e engineering as gender discrimination. Gender discrimination is, in Bob Brown's language, a problem that can be solved, and he and others at MIT have begun to address it vigorously. The departments with the worst records of hiring women have been served notice that they must do much better, starting now. Already the difference in hiring practices is discernible, as is the difference in filling leadership positions. The new executive officer (roughly equivalent to an associate department head) of the Department of Chemical Engineering is a women, as is one of the newly appointed associate heads of the Department of Electrical Engineering and Computer Science. MIT's new vice president for research is a woman chemical engineer.

Despite the recent changes, MIT remains a profoundly masculine place. It has done a lot to reduce gender discrimination in engineering, but it will be a much longer and more difficult task for MIT to address the gender bias of engineering. Discrimination has to do with obstacles to full participation in an enterprise; bias has to do with the cultural identity of the enterprise. Gender bias arises from the network of behaviors—intellectual, cultural, artistic, social, psychological—that link the sexual dualism of human beings with the full range of human activities. Gender bias is a fact of human life, not necessarily a problem to be solved. It becomes a problem when the linkages of gender bias form social structures that routinely damage people (men, women, or both) and the lifeworld.

Gender discrimination and gender bias are deeply interconnected. The cultural identity of science and engineering as masculine enterprises results from, and feeds back into, the overwhelming prevalence of men in their practice. As gender discrimination is gradually addressed, gender bias should also change. That is the point of increasing diversity in any human institution: not only to ensure equal access to the advantages it offers, but also to redefine the institution so that its identity reflects the full range of human needs and concerns. One reason to open up engineering to women is to be fair to women. Another reason, ultimately more important, is to transform engineering so that it better serves the needs of humanity. Women want a life, just as men do. And women truly want to make a difference, just as men do. It is not going too far to say that the future of democracy depends on whether we can be as inventive socially as we are technologically in devising ways for people to have a life while also making a difference.

## How Gender Bias Works

In a democratic Western society, it is relatively easy to achieve consensus on the propositions that gender discrimination should end and that we should be fair to women. It is harder to achieve consensus on the idea that gender bias should end. There the argument shifts from equality to identity, and specifically to essentialism: the belief that women practice engineering differently because they are different from men. This kind of argument makes many women as well as men uncomfortable. It is reminiscent of biological determinism and of biology-based arguments that have been used to justify discrimination against women. Identity cannot be reduced to biology. Women's identity comes overwhelmingly from socialization, and the identity of an engineer comes entirely from socialization.

No one has useful evidence about essential differences; our only evidence comes from observed differences. In the world of engineering as it now exists, the most obvious observed differences between men and women are not in ability, not in standards of excellence, but in what interests them. One striking example is the difference between male and female choices of undergraduate engineering majors. Women strongly prefer chemical engineering (as noted, 62 percent of the undergraduate majors at MIT are women), which is now oriented to the needs of biotechnology. They prefer civil and environmental engineering, primarily its environmental wing, even more strongly (75 percent of the undergraduate majors). They also tend, although somewhat less strongly, to choose aeronautics and astronautics as a major, especially since that department has strongly emphasized team-based education and a holistic systems approach to engineering.

In contrast, women make up only about 33 percent of undergraduate majors in mechanical engineering and 27 percent in electrical engineering and computer science. Since these two departments have particularly poor records in hiring and keeping women faculty (the 54-person faculty of the Department of Mechanical Engineering includes two women), their absence may deter young women from majoring in these fields. National figures, however, support the preference of women for branches of engineering that focus on environmental and biological problems. There is a similar lopsidedness in enrollments in the School of Science, where women overwhelmingly prefer biology, environmental science, and neuroscience to physics or math. (More than three-fourths of the majors in Earth, Atmospheric, and Planetary Sciences are female.)

Are these preferences of interest within engineering problematic or not? Should the goal be to ensure that as many women go into mechanical engineering as into bioengineering, for example? What is a truly free choice, as opposed to one guided by subtle social signals? These are the kinds of questions that MIT, and engineering more generally, must face now that the initial and most obvious barriers of gender discrimination have been breached. Gender bias persists but is now being played out on a finer level of detail. No longer is the stereotype true that science and engineering are for men only, but now science and engineering are being sorted out into male and female domains. The stereotypes are discussed openly by undergraduates: EECS is the tough, rigorous major for men; biology and "environmental stuff" are for women.

This pattern of finer-grained gender bias is evident in other professions that have recently seen an influx of women. In medicine, surgery is dominated by men, pediatrics by women. In law, men dominate criminal and corporate law; women more often handle domestic litigation. It is not a problem that women have different preferences in professional life than men. What is a deep and structural problem is that professional areas more often chosen by women bring smaller financial rewards and offer fewer opportunities for advancement than those typically preferred by men.

This pattern is being reproduced in the realm of engineering, but with an important twist: the structural differences in reward and recognition are now not so much among the branches of engineering as between engineering as a whole and the more expansive and more lucrative realm of "technology." In many ways, engineering is being feminized. In engineering education, communication and other "soft" skills are getting more attention in the curriculum and in student consciousness. More time is spent on teamwork, role playing, and negotiation. Even in solving "hard" technical problems, engineering students are told they must be "sensitive" to the larger context.

There was always a housewifely role in engineering.[4] I heard it when my male relatives griped about their essential, unglamorous, and unappreciated labor. That role is becoming more and more important as the management of a reflexive, hybrid world requires forms of education, consciousness, knowledge, and skills traditionally associated with femininity. This shift in the gender bias of engineering is happening not because more women are going into engineering (it is hardly a flood) but because of the objective requirements of the world in which we live.

But even as the rhetoric of engineering softens, the rhetoric of technology is getting harder. Maybe women get good grades, but can they handle the rough and tumble of entrepreneurial creativity? The ideology of technological change is full of tough-guy rhetoric about the need to stay out front, to beat the competition, to work 24/7 to keep ahead in a brutal environment. Not a world for girls. Of course, no one ever says directly that women are excluded from technological innovation, but the emphasis on the constant, brutal demands of the high-tech workplace implies once again that they have to make a choice between work and family. Technology and not engineering, innovation and not upkeep—these are where the serious money and power lie, and these realms are represented as male and controlled by men.

But it gets even stranger, now that "technology" and "innovation" are joined in our culture. The glorification of technology as innovation becomes quasi-mystical praise of male procreation. Innovation, the endlessly creative process of which men are in charge, promiscuously generates new life forms to fill the earth. Men are the ones who endow new machines with a soul.[5]

Captains of industry have been replaced by captains of innovation, the male procreators of change. Innovation is a life force,
endlessly fertile, filling institutional incubators that nurture its
infancy. The ultimate dream is "the engineering of life."[6]

To say that these odes to innovation have a hysterical edge
just reinforces the point. Language, again, gives it away.
Women do not have to be so shrill in insisting on change. They
embody it. "Conceive, design, implement, operate." I halt at the
first word and try to imagine how any woman could have come
up with this sequence of imperatives. In this male version, conception is defined as a strange form of fertility, which ends in
operating a device a man has engendered. Other forms of innovation (artistic, social, intellectual) are ignored.[7] The fact that an
increasing number of women participate in engineering is both
fine and irrelevant, if the cult of innovation defines a distinctive
technological arena where men and masculinity dominate. The
cult of technological creativity leads to some weird rhetorical
cross-dressing, as fertility is claimed as a male prerogative in
defiance of long-standing cultural understandings.

This rhetorical mess points to the deeper problem of gender
associations in the human-built world. Technological activities
tagged as masculine have played the dominant role in producing that hybrid world. They are less suited for organizing a
technological world we can live in. For this, "soft" engineering
skills are increasingly necessary. Many male engineers are not
especially adept at such skills, and they are not necessarily the
skills that bring prestige and power in our society.

MIT is trying to reconcile the goals of change and community.
The danger is that gender bias will increasingly identify
"change" as a male role and "community" as a female one. The
rhetoric of change encourages this to the extent that change is
described as aggressive, individualistic, and all-consuming. More

than rhetoric is involved, however. There is measurable evidence that women do more than their share of community service at the Institute. The School of Engineering report on women includes two plots of outside professional activities, one for compensated activities and one for uncompensated activities, broken down by gender. Men perform significantly more compensated activities, women significantly more uncompensated ones. Both this report and the earlier one on the status of women in the School of Science show that in some departments women are routinely ignored when it comes to serving on committees that act with power in academic matters (search committees, for example). They are, however, frequently asked to serve on committees related to maintenance of the academic commons.

These measurable differences in male and female behavior imply differences in socialization and priorities that work to the disadvantage of women in a world where change is done for pay and "community" is done pro bono. Organizing a technological world we can live in requires change and community as inseparable processes, not as gender-typed opposites. We can never succeed in this if innovation is defined as a technology-intensive, male-dominated role, while women are left to deal with the consequences of innovation. "Building community" also requires entrepreneurship and innovation. Change should not be identified only with technology when our deepest needs for innovation also require changes in social organization.

## Retooling Gender Roles

Gender bias in technology and engineering is not a "feminist issue," not peripheral, not a special interest. It is central to our ability to organize a technological world we can live in. The principles for the organization of that world must be co-evolved by men and women; otherwise we will have a human-built world that is not a humane one.

Today's interplay of changes in technology and gender roles is another reminder why the Neolithic Revolution is the most compelling precedent for the technological revolution of our times. In pre-agricultural societies, the female role combined production of new human life and the collection of plant food, while the male role was dominated by hunting game and defending the group. It is plausible to assume that these roles were universal, or nearly so, before the invention of agriculture, and that this invention started a process of profound change in gender roles. Once humanity became "bound to the soil," men were largely deprived of their traditional role as hunters and claimed new roles as cultivators—both in a literal sense (assuming primary responsibility for plant cultivation, which had been women's work) and in a metaphorical sense (claiming primary responsibility for procreation, as sowers of the seeds of life). Women's role in food production was diminished, while their role as bearers of children (having passively received the male-sown "seeds") became more significant, since farming benefits from a growing population. At the dawn of recorded history, the Neolithic Revolution transformed production, reproduction, and the cultural understanding of both.[8]

Another set of interactive changes are taking place today. The most obvious technological change affecting gender relations is the multiplication of reproductive choices due to new methods of fertilization, cloning, and the like. Among other possibilities, these methods allow the extension of fertility beyond its "natural" limits. As in Neolithic times, technological innovation brings with it new reproductive facts of life for both genders. What is means to *father* a child inevitably changes along with new options for childbearing. The emphasis on male procreation of technology seems to get more intense as the male role in biological procreation becomes increasingly uncertain.

The new facts of life today, however, extend beyond biotechnological ones. The transition from a world dominated by nature to one dominated by technology creates other new facts of life, which again affect men as much as women. In particular, this transition deprives men of many of their now-traditional working roles. The male productive role is at least as much changed in the transition from an industrial to an information society as it was in the transition from a hunting to an agricultural society.

What does it mean to be a man in a technological world? We are so used to associating technology and engineering with men that it takes some effort to realize how unlikely, even strange, that association is in the new hybrid world. The cultural patterns we are accustomed to do not well fit the material patterns of the world we are constructing. The advantages of masculinity in labor are not evident in the information age, when labor involves manipulating keyboards and buttons, reading, writing, calculating, and speaking. Even in social violence, the male advantage declines when the means of violence become highly technological. There are still opportunities for physical heroism in combat and in rescue operations, but even in these areas technological development is directed toward limiting the number of men on the ground, limiting the physical risk and toil, limiting the possibility of harm. In movies and television shows, when future heroism is portrayed, men are still usually the heroes—but their resoluteness is often shown as punching in the correct code on a computer keyboard.

From this perspective, the problems of women in creating a livable life at MIT, and in other social settings dominated by science and technology, seem all the more astonishing and regrettable. In many ways, women should have an easier time adapting to the new hybrid world, because their socialization is generally congruent with the requirements of these complicated and profoundly humanized surroundings. It is much less obvious how men will find an appropriate masculine identity in a human-built world as physical strength becomes less and less important in ordering that world. Defining a place for women in engineering is a serious issue. The deeper, more perplexing issue is defining a place for men in a technological world.

# 6

# Coda: Living in a Historical World

I had planned to end the book here.[1] I was not entirely happy with this conclusion, preferring one that would connect more strongly with the beginning. The original ending did not circle back to my opening dream, to Spring Garden Farm, or to my grandfather's unintentional participation in the destruction of a world he loved. My cousin Mary, who helped me edit the first chapter, suggested that my use of the word 'destruction' in it was too strong. I agreed and reworded the passage to say that my grandfather had contributed to the erosion of his world. Mary also said that she especially liked my phrase about our grandfather's being "hijacked by the twentieth century," and we discussed it as a possible title for the book.

Shortly after our conversations, I stopped on my way to work to have my car serviced. At the Acura dealership, I watched one tower of the World Trade Center burn, then saw a jet fly into the second tower. It took us a while, though not long, to learn that both planes had recently taken off from Logan Airport, only a few miles away. Those of us who happened to be there that day will never return to the Acura dealership without flashbacks. History, and the binding power of historical events, has reasserted itself.

If this book began with a dream, it ends with a nightmare. 'Destruction' no longer seemed too strong a word. A title about hijacking was out of the question. There have been times when I have wished, or even halfway imagined, that Grandpa Lewis could see me and my family and MIT to take pride in it all. This time I felt relieved that he could not see.

I reread the manuscript and changed only a few words. I felt I had already addressed most of what I needed to say about the events of September 11, 2001. I decided, however, to write this coda in order to underline some of what has been said in preceding pages.

Disasters are revelations. With unforgiving clarity, they show what is going on "normally." We never understand a technological system better than when it collapses. The process of destruction unmasks design flaws, and so technological disasters are followed by technological post-mortems. We peer into the ruins to figure out what things need fixing: the O-ring on the Space Shuttle, the cooling system of the nuclear reactor, the building struts, the cockpit door.

But when a technological disaster is caused by deliberate human action, when "normal" civilian technologies are turned into weapons, we are forced to think more deeply about technological systems—not only about their material design but also about their meaning. In nightmare images, the events of September 11 revealed the existential reality of the reflexive nature of the technological world: the airplane turned into a missile, the building creating its own implosion, the Pentagon transformed into a target rather than a command center.

It is bad enough to realize that such gigantic and powerful constructions can so readily be turned against their creators. It is worse to realize that the familiar, unprepossessing systems of daily life can so easily be turned against their creators: "How easy it is to enter the country and commit a massive act of terror by using the tools of everyday life, such as easy access to the Internet, convenient banking and credit, and many ways to travel quickly."[2] Even the mail system, one of the oldest and apparently most benign technological systems of civilization, can easily be turned into a network of bioterrorism.

Modern science emerged from a belief that design in nature reveals God's purposes. In a hybrid world, technological design reveals human purposes. The technological world, as it has been designed in places like MIT, serves purposes of instantaneous communication, rapid and frequent travel, intellectual and emotional stimulation, convenience, and comfort. It can be criticized as commercial and consumerist, but it is often engaging and enlightening. Whatever its virtues and vices, it is the world we are used to.

Now we peer into this familiar world and see in its depths a frightening one, which we have been constructing all along with only the faintest awareness of what we were doing. At an MIT teach-in on Technology, Terrorism, and War held on October 1, 2001, much of the discussion centered on the "developmental divide" that suddenly seemed more significant than the more narrowly defined "digital divide." One panelist—David Marks, a professor of civil engineering—described the material, social, and political dangers of the megacities of the developing world. These are places like Bangkok or Calcutta, where 20 million people live in dense agglomerations (tinderboxes, Marks called them) that are inverse, impoverished versions of the human concentration of the World Trade Center. The question-and-answer part of the teach-in revealed a new awareness that the technological world has built into its design the creation of social marginalization: unemployment, disease, poverty. As someone remarked, this too is globalization.

I thought about these tinderboxes as the outcome of humanity's collective decision to unbind itself from the soil—a decision that has led humanity to dwell in urban squalor far more often than in virtual space. I also began to think about divides in the technological world even deeper than that between developed and undeveloped societies. One of those divides is that between male and female. The strongest motivations that led the terrorists to destroy technosocial systems identified with the West did not arise from outrage against social injustice. They were motivated by religious beliefs in which fear of and hatred for women are not incidental but central elements. The terrorists were uprooted men who found identity and meaning by asserting superiority over women and unbelievers through technological prowess. For all their denunciations of Western values, they were gripped by a Western fascination with technological display. The transition to a world dominated by technology, not nature, deprives many men of obvious productive roles. It also multiplies reproductive choices and productive roles for women. In such a world, the surest way to reassert male identity is through a return to arms.

"We have to catch up with the impact we've had on the world." That impact is at least as much psychological as it is material, economic, and social. In the psychological realm we are only beginning to come to terms with the consequences of our actions. We cannot keep generating innovations without giving much more attention to our ability to live with the changes we generate. This means not a retreat from innovation but a redirection of it toward the central challenge of our times: "to organize a technological world we can live in."

The attack on the World Trade Center was an unforgettable and in its own way brilliant image of hatred and destruction. It also provides an equally unforgettable image of a world we could live in. In the imagination of disaster, catastrophe can have the paradoxical effect of restoring community. It reveals the grandeur latent in individuals and societies; it sweeps away conventional social distinctions; it reaffirms social bonds eroded by the relentless corrosion of "change." That is because the collapse of the material elements of society reveals the essential human elements that endure. When technology falls apart, what is left?

What is left, first, is history. In the years before September 11, 2001, it had seemed that technology had displaced history as the comprehensive process of change over time. At MIT and in many other places, we were told over and over again about the inevitability of technological change and the need for change management by change agents. Change was a new software system, a new product, a new technique. Technology, not history, was going to shape the future.

In the months after September 11, 2001, we were told over and over again that everything had changed. What has changed most of all is our understanding of change. We know history when we see it. We know we are experiencing change that is significant, that is historical—historical because it involves human relationships, meanings, and expectations.

At the MIT teach-in, John Hansman commented: "In twenty minutes on September 11 our view of the civil aviation system changed completely." The technological details of civil aviation did not change at all in those minutes, but everything else did, and John Hansman understood that the non-technological changes had altered the entire system. The alteration in meaning will have complicated effects on how people travel, whether they travel, how the system is operated, and, over time, how the system is designed. In the sequence of "conceive, design, implement, operate," conception is the decisive step.

The other thing that is left when the material part of technology collapses is humanity. We always knew that technological systems are composed of both material and social elements, but, as they saying goes, now we get it. That is why the technological catastrophe was also a human catastrophe. People died because all the interlocking systems—aviation, military, safety, health, information—were crawling with humanity: passengers on airplanes, emergency workers in streets, knowledge workers at desks, medics in ambulances, security checkers in airports, mail sorters, postal carriers. They were men and women of all colors, nationalities, languages, and levels of education, only a few of whom could be called engineers. All of them had their lives bound up with the creation, the maintenance, and the use of technological systems.

When the material part of the systems disintegrated, human beings rose to the occasion: firefighters charging to the rescue, husbands and wives phoning parting messages of love, passengers on one airplane resolving to die rather than surrender. They were heroic, but the revelation here is the heroism of everyday life, of people doing their jobs, of people loving families and friends. The social and personal connections are always entangled with (and sometimes obscured by) the technological connections, but ultimately they are revealed as the essential ones. In short, disaster revealed the core truth of technology and science studies: that technoscience is embedded in human history and human society.

On September 13, 2001, the MIT administration cancelled classes after 3 P.M. and invited the community to gather in Killian Court, the great expanse of grass, framed by elm trees and rhododendrons, that rolls from the foot of the Great Dome to the Charles River. At a few minutes before 3 o'clock, people began coming into Killian Court from the buildings that surround it on three sides. They kept coming and coming—staff, students, and faculty members—streams swelling into a human sea. At the hour, a recording of requiem music that had been performed in Prague the year before by the MIT student orchestra was played.

After a few brief remarks from the organizers, the thousands of people gathered in the court divided into groups of about fifteen, each gathered around a staff or faculty person who had volunteered to lead a discussion. It took only a few seconds for my group to form. For more than an hour we sat on the lawn, talking. It was a magnificent afternoon, late in the day and late in the season, dry and calm. The sun, low in the sky, illuminated the names of the great scientists engraved on the facades around Killian Court. We sat at a place on the lawn where, in my former life, I had often gazed as I daydreamed during Reengineering meetings.

If wisdom begins with the acknowledgment of one's ignorance, that day we began a quest for wisdom. My group consisted almost entirely of students, graduate and undergraduate. We talked about how the terrorist attacks had affected us, how we were responding, how we planned to respond. No one was especially interested in the technical details of how this had happened or even in how it could be prevented from happening again. The dominant question was why it had happened. Many students said that this disaster revealed to them how little they really knew about the world, despite their computer literacy and their immersion in so-called information. Most of those who had been raised in the United States could not understand why anyone could be so angry at their country. Several students from other countries said that, while they too were horrified, they thought they understood. Some of them described similar experiences of terror in their homelands and explained how they had learned to live with fear.

I thought of Keats's lines describing how he felt when he first read the poetry of Homer in Chapman's translation:

Then felt I like some watcher of the skies
When a new planet swims into his ken

For us, that new planet was our own earth. But we also talked a lot about our inner worlds—the mixtures of feelings and thoughts we were experiencing, which we had also discovered with (to quote again from Keats) "wild surmise." Our conversation was about knowledge, of the world and also of the self. Several students said they were rethinking what they wanted to do with their lives.

This, too, is what disaster reveals: how important education is, and what education should be like, in our time. That afternoon, in Killian Court, the humanities, the arts, and the social sciences were at the very core of MIT education. "Education is preparation for life." We were beginning to prepare for life in a hybrid, reflexive world. History, literature, art, and anthropology never seemed more relevant to the lives of scientists and engineers. They never seemed more important to the lives of everyone.

But the most important element of the educational experience that day lay in the diversity of the students gathered on the lawn. Students who saw the world from the vantage point of the United States learned from the students who saw the world from other vantage points. We did not just exchange views and then log off. We all knew that we would continue to live and learn together, in the same classrooms and labs and maybe even in the same residences, for days and weeks and months to come. A few days later, I heard that guards had been posted around the Muslim prayer room in the MIT Religious

Life Center to make sure that no one was threatened or harmed there. The center had been designed to bring together those who practice different faiths so they are immediately aware of one another on a daily basis. The Muslim prayer room is located next to the room where Jewish students worship. This concentration of human diversity in Cambridge 02139 is difficult to manage. This very difficulty is why it is so valuable.

I hope to teach at MIT for many more years, but I do not think I will ever have an hour that so well expresses the quintessence of education as that transcendent hour in Killian Court. That afternoon, MIT became a university. That afternoon, for the first time, I felt as a teacher what I had never quite felt as an administrator: that I belonged at MIT, that I was just as much a part of it as my grandfather.

# Notes

## Chapter 1

1. As Leo Marx has shown in *The Machine in the Garden: Technology and the Pastoral Ideal in America* (Oxford University Press, 1964, 2000), we do some of our deepest thinking about technology in images. For Americans, a machine invading a rural retreat is a recurring image: Nathaniel Hawthorne hearing the shriek of the locomotive whistle as he rambles in the neighborhood of Concord on a summer Sunday morning, Henry David Thoreau listening to the rumbling of the "iron horse" in his Walden retreat, the fictional Huck Finn seeing a steamboat emerging from the fog to smash the raft on which he and Jim are floating down the Mississippi. When the Dual cut across Spring Garden Farm 50 years ago, it fit this classic image of technological change in America—the machine invading the garden. Other images are emerging to convey the all-encompassing character of the technological landscape, where the emblematic sound of the human-built world is not an occasional whistle but the "white noise" of ceaseless traffic. (See Don DeLillo's novel *White Noise*, published by Penguin in 1984.)

2. My title changed and my responsibilities expanded during the 5 years I served. For the last 4 years of my tenure, my title was Dean of Students and Undergraduate Education, and the following units reported to me: Academic Services (including Registrarial Services); Admissions; Athletics, Recreation, and Physical Education; Campus Activities Complex (including the Office of Campus Dining); Career Services and Preprofessional Advising; Counseling and Support Services; Minority Education; Residential Life and Student Life

Programs; and Student Financial Services. I resigned as dean effective July 1, 2000, in order to return to the faculty.

3. I am indebted to Lydia Snover, Assistant to the Provost for Institutional Research, for providing up-to-date information (accurate as of March 2002) on MIT's faculty and students.

4. See Nicholas Negroponte, *Being Digital* (Knopf, 1995). (Negroponte, the longtime director of MIT's Media Lab, recently stepped down from that position.) The dichotomy of lifeworld and systems is associated with Jürgen Habermas; see Fred R. Dallmayr, "Life-World: Variations on a Theme," in *Life-World and Politics: Between Modernity and Postmodernity*, ed. S. White (University of Notre Dame Press, 1989), pp. 45–46 and notes. The dialectic of globalization and identity is a major theme of Manuel Castells, whose works are referenced throughout this book. The term "symbolic analyst" comes from Robert Reich, *The Work of Nations: Preparing Ourselves for 21st Century Capitalism* (Knopf, 1991). The quote about the disposable labor force is from Manuel Castells, *The Rise of the Network Society*, second edition (Blackwell, 2000 [1995]), p. 295.

5. This discussion is based on a book by Edward T. Layton Jr.: *The Revolt of the Engineers: Social Responsibility and the American Engineering Profession* (Press of Case Western Reserve University, 1971), p. 53. The historian of technology Susan Douglas reminds us that "the study of ideology is essential to the study of technology." She defines it as "the entire constellation of images, myths and ideas through which people come to understand the material world and their place in that world." Social groups are engaged in an ongoing struggle about the myths and meanings that will be accepted as the natural and obvious way to interpret social reality. "This," Douglas claims, "is a struggle invariably won by the dominant classes." (Susan J. Douglas, "Jürgen Habermas Meets Mel Kranzberg: What Media Theory Has to Offer the History of Technology and Vice Versa," paper presented at annual meeting of Society of the History of Technology, 1987, p. 2.)

6. Karl Mannheim, *Ideology and Utopia: An Introduction to the Sociology of Knowledge* (Harcourt, Brace, 1936 [1929]), p. 203; see pp. 192–211 passim.

7. In listening for these words, their clustering, and their mutual defining, I am drawing on an honorable tradition in intellectual

history—best exemplified by Raymond Williams's book *Culture and Society, 1780–1950* (Columbia University Press, 1958)—of analyzing the social and cultural implications of the industrial revolution. Williams highlighted the reflexive or circular relationship between key terms of social analysis in the nineteenth century (culture, society, industry, art, etc.) and the changes they supposedly described. In addition, Williams showed how key concepts tended to cluster, so that their meanings shifted in response to alterations in the others. See Marx, "Technology: The Emergence of a Hazardous Concept," *Social Research* 64 (1997), no. 3, p. 967.

8. Leo Marx, who has tracked the use of the word, calls 'technology' a "hazardous concept," unstable and reflexive in meaning, and particularly dangerous when used as the subject of a sentence, implying that it is an independent, autonomous historical agent, as in "the archetypal sentence: 'Technology is changing the way we live.'" Marx argues that the word 'technology' began to be used in the late nineteenth century to fill a "conceptual void" in explaining how the utilitarian, instrumentalist branch of human activity, directed at controlling the physical world, could so powerfully generate social change. (Marx, "Technology: The Emergence of a Hazardous Concept," pp. 968, 974–975, 977.) The naming of MIT in 1862 was one of the first public uses of the term. (The choice probably was influenced by Jacob Bigelow, a Boston botanist and physician, who had used the term in a Harvard lecture in 1828; he later became an MIT trustee.)

9. John Staudenmaier, "Perils of Progress Talk: Some Historical Considerations," in *Science, Technology, and Social Progress*, ed. S. Goldman (Lehigh University Press, 1989).

10. The process by which technological change began to displace historical change began in the seventeenth century with the controversy of the "ancients" and the "moderns." The "moderns" might concede equality or even superiority to the ancients in arts or philosophy, but they felt they clinched their case in arguing for progress when they emphasized that only the moderns possessed the compass, the printing press, gunpowder, and (most important) the experimental method. These arguments from progress in inventiveness were extrapolated to history as a whole; progress in what we now would call technology was cited as irrefutable evidence for historical progress from antiquity to the modern age. When technology entered the study of history, history itself began to be redefined as the study

of the record of human progress. See Richard Foster Jones, *Ancients and Moderns: A Study of the Background of The Battle of the Books* (Washington University Studies, New Series in Language and Literature, no. 6, 1936). See also Hannah Arendt, *The Human Condition* (Doubleday Anchor Books, 1959 [1958]), esp. pp. 119–133.

11. Anthony Giddens, *The Consequences of Modernity* (Stanford University Press, 1990), p. 51. Giddens puts "emptying out" in quotes in the text I am paraphrasing.

12. As Douglas reminds us in "Jürgen Habermas Meets Mel Kranzberg," the "'engineering of consent' is not a one-time event, in which meanings are clamped down and everyone accepts them." Consent is by no means automatic. It is an ongoing process in which people adjust their ideas and beliefs but also protest in a variety of ways. If the ideological framework becomes too much at odds with daily experience and with common sense, then the disparity has to be resolved for the social order to be stable: "The dominant ideology must incorporate certain criticisms and concerns while successfully marginalizing others, and illustrate that such concerns can be resolved best under the existing sociopolitical system."

13. Thomas Friedman makes this point when he uses a Lexus automobile and an olive tree to represent two sides of a technological divide: "Half the world seemed to be emerging from the Cold War intent on building a better Lexus, dedicated to modernizing, streamlining and privatizing their economies in order to thrive in the system of globalization. And half of the world—sometimes half the same country, sometimes half the same person—was still caught up in the fight over who owns which olive tree." (*The Lexus and the Olive Tree*, Farrar, Straus and Giroux, 1999, p. 27.)

14. Alex Roland, email, March 5, 2000.

15. See Tom Standage, *The Victorian Internet: The Remarkable Story of the Telegraph and the Nineteenth Century's On-Line Pioneers* (Walker, 1998).

16. The concept of revolution migrated physics (the path of the planets around the sun) to history (the overturning of British rule in America in the 1770s, the end of the monarchy in France in 1789), and then, in the early nineteenth century, in a deliberate analogy with events in France, was applied to machine production in the metaphor of the industrial revolution. When the metaphor of revolution is applied to industry, it implies a suddenness and completeness that

might occur in the political arena (the king is executed, Robespierre rules) but does not apply in the arena of production. But to say that the metaphor is misleading is to miss the point. In the words of Leo Marx: "The whole issue becomes irrelevant once we recognize that we are dealing with a metaphor, and that its immense appeal rests, not on its capacity to describe the actual character of industrialization, but rather on its vivid suggestiveness. It evokes the uniqueness of the new way of life, as experienced, and, most important, it is a vivid expression of the affinity between technology and the great political revolution of modern times." (*The Machine in the Garden*, p. 187n)

17. Steve Lohr, "The Future Came Faster in the Old Days," *New York Times*, October 5, 1997.

18. Lewis Mumford calls the invention of language "infinitely more complex and sophisticated . . . than the Egyptian or Mesopotamian kit of tools" ("Technics and the Nature of Man," *Technology and Culture* 7, 1966, no. 3, p. 308).

19. On the term 'habitat', see Gayle L. Ormiston, "Introduction," in *From Artifact to Habitat: Studies in the Critical Engagement of Technology*, ed. G. Ormiston (Lehigh University Press, 1990), p. 13.

20. Elting Morison, "Introductory Observations," in Morison, *Men, Machines, and Modern Times* (MIT Press, 1966), p. 16. See my discussion of the language and imagery of the artificial environment in *Notes on the Underground: An Essay on Technology, Society, and the Imagination* (MIT Press, 1990), pp. 1–8.

21. Manuel Castells (*The Rise of the Network Society*, p. 15) puts it into one long sentence: "Matter includes nature, human-modified nature, human-produced nature, and human nature itself, the labors of history forcing us to move away from the classic distinction between humankind and nature, since millenniums of human action have incorporated the natural environment into society, making us, materially and symbolically, an inseparable part of this environment."

22. Bill McKibben makes this point in *The End of Nature* (Random House, 1989).

23. Ulrich Beck, *Risk Society: Towards a New Modernity* (Sage, 1992 [1986]), p. 81.

24. Here I am paraphrasing Mumford ("Technics and the Nature of Man," p. 303), who writes: "In terms of the currently accepted

picture of the relation of man to technics, our age is passing from the primeval state of man, marked by his invention of tools and weapons for the purpose of achieving mastery over the forces of nature, to a radically different condition, in which he will not only have conquered nature but detached himself completely from the organic habitat."

25. Mumford, "Technics and the Nature of Man," p. 303.

26. See Stanley H. Ambrose, "Paleolithic Technology and Human Evolution," *Science* 291 (2001), no. 5509, pp. 1748–1752: "Stone tool technology, robust australopithecines, and the genus *Homo* appeared almost simultaneously 2.5 [million years ago]." (p. 1748) The systematic use of fire evidently began between 1 million and 1.5 million years ago, and around 300,000 years ago "technological and cultural evolution accelerated" in a way that suggests "the emergence of true cultural traditions and cultural areas" (p. 1751). The stone tool technologies between 2.5 and 0.3 million years ago "are remarkable for their slow pace of progress . . . and for limited mobility and regional interaction. . . . With the appearance of near-modern brain size, anatomy, and perhaps of grammatical language ~0.3 [million years ago], the pace quickens exponentially." (p. 1752)

27. See Peter F. Drucker, "The First Technological Revolution and Its Lessons," *Technology and Culture* 7 (1966), no. 2, pp. 143–151. This article is Drucker's presidential address to the Society for the History of Technology, presented on December 29, 1965, in San Francisco. He defines "the first technological revolution" as "the irrigation city, which then rapidly became the irrigation empire." (p. 143)

28. Mircea Eliade, *The Forge and the Crucible*, second edition (University of Chicago Press, 1978 [1956]), pp. 177–178. For a discussion of this passage in a different context, see Williams, *Notes on the Underground*, pp. 1–4.

29. The concept of reflexive modernization was articulated most clearly by Beck in *Risk Society*. See also Anthony Giddens's books *The Consequences of Modernity* (cited above) and *Modernity and Self-Identity in the Late Modern Age* (Stanford University Press, 1992).

30. Giddens, *The Consequences of Modernity*, p. 38.

31. Beck, *Risk Society*, pp. 12–13.

32. Under the governing structures of market relationships, the abiding imperative of capitalist accumulation—to increase profits—is accomplished in a few major ways. The circulation of capital can be

maximized through increasing its extension in space (broadening the market) and its intensity over time (accelerating capital turnover). Also, production costs can be lowered, productivity can be raised, and new products can be brought to market. Particularly in the last three decades, all these ways of increasing profit are closely linked to technological innovation, which has become the primary driver of capitalist acquisition. See Castells, *The Rise of the Network Society*, pp. 68–69, 95. Castells comments: "Information processing is focused on improving the technology of information processing as a source of productivity, in a virtuous circle of interaction between the knowledge sources of technology and the application of technology to improve knowledge generation and information processing. . . . Or, more briefly, a prime characteristic of the information age is its circularity, characterized by the immediate application to its own development of technologies it generates. . . ." (ibid., p. 32)

33. "Meanwhile man . . . exalts himself to the posture of lord of the earth. In this way the impression comes to prevail that everything man encounters exists only insofar as it is his construct. This illusion gives rise in turn to one final delusion: It seems as though man everywhere and always encounters only himself. Heisenberg has with complete correctness pointed out that the real must present itself to contemporary man in this way. *In truth, however, precisely nowhere does man today any longer encounter himself, i.e. his essence.*" (Martin Heidegger, "The Question Concerning Technology," in *The Question Concerning Technology and Other Essays*, Harper & Row, 1977, p. 27) Hannah Arendt (*The Human Condition*, p. 237) quotes Heisenberg's observation about "man encountering only himself."

34. The first quote (from Morison, *From Know-How to Nowhere*) is cited by Thomas P. Hughes in "Memorials: Elting Morison, 1909–1995," *Technology and Culture* 37 (1996), no. 4, p. 874. The second quote, also from "Memorials," is cited by Leo Marx (ibid., p. 867).

35. Manuel Castells, *The Power of Identity* (Blackwell, 1997), p. 8.

## Chapter 2

1. Nancy Hopkins, "MIT Report on the Status of Women Faculty in Science Leads to New Initiatives to Increase Faculty Diversity," *MIT Faculty Newsletter* 13 (2000), no. 2, p. 15.

2. Joseph A. Schumpeter, *Capitalism, Socialism and Democracy*, third edition (Harper & Row, 1950 [1942]), pp. 81–86.

3. Theodore Caplow, Louis Hicks, and Ben J. Wattenberg, *The First Measured Century: An Illustrated Guide to Trends in America, 1900–2000* (AEI Press), as reported in the *New York Times* on December 2, 2000.

4. Remarks of Bob Whitman, "Tributes to Charles Miller from former students, colleagues and co-workers at memorial service," *Civil and Environmental Engineering at MIT* 15 (2000), no. 1, p. 8.

5. The Department of Civil and Sanitary Engineering, formed in 1934, changed its name to Civil Engineering in 1961.

6. "Innovations in University/Industry Alliances: An Interview with Karl Koster," *MIT Report* 29 (2001), no. 3, p. 2.

7. I am not vouching for the reliability of these numbers. They are from typewritten manuscripts of talks my grandfather gave on topics like "the engineering profession" and "the place of engineering in society and civilization," which I have in my possession.

8. Years later, when I began to study the history of technology, I realized that these conversations opened a door for me not only into the real world of organizations and responsibility but also, and more specifically, into the collective consciousness of modern engineering. In Layton's terms, they represented efforts to reconcile the ambiguous position of the engineer in a capitalist economy. See Layton, *The Revolt of the Engineers*. See also David Noble, *America by Design: Science, Technology, and the Rise of Corporate Capitalism* (Oxford University Press, 1977), esp. pp. 35–39. For a more recent work, see Robert Zussman, *Mechanics of the Middle Class: Work and Politics Among American Engineers* (University of California Press, 1985), esp. pp. 4–9.

9. Layton, *Revolt of the Engineers*, p. 73. See pp. 53–74, passim.

10. Ibid., p. 54.

11. Ibid., p. 3. The expansion of the engineering profession to include more and more of "technology" happened only in the nineteenth century, as engineers began to move into industry in large numbers. Layton tells us: "In 1816, the engineering profession scarcely existed in America. It has been estimated that there were only about thirty engineers or quasi-engineers then available; but by 1850, when the census first took note of this new profession, there were 2,000 civil engineers." The expansion became even faster after 1850, when the

profession expanded beyond its base in civil engineering to encompass technical specialists who worked to meet industrial needs, first in a shop and later largely in a corporate setting (mechanical, mining, metallurgical, electrical and chemical engineers). Most of this expansion took place in the newer and most scientifically influenced fields of chemical and electrical engineering. The rate of expansion slowed somewhat during the Depression, but picked up again after World War II: 500,000 engineers in 1950, more than 800,000 in 1960, and more than 1.5 million in 1981.

12. Ibid., p. 9. The study was conducted by the Society for the Promotion of Engineering Education.

13. Robert K. Weatherall, "A New Breed of Engineering Student," *Careers and the Engineer*, August 1989.

14. Robert K. Weatherall, Engineers as Supervisors and Managers, unpublished manuscript, based on data in the National Science Foundation's Scientist and Engineers Statistical Data System (March 1999), p. 4, drawing on statistics in William K. LeBold, Robert Perrucci, and Warren Howland, "The Engineer in Industry and Government," *Journal of Engineering Education* 56/57 (1966), March, p. 259 (appendix B).

15. One of the most common themes in the history of technology is the nature of the relationship between science and technology. American historians of technology, including Layton, tend to emphasize how science and technology have been distinct activities, even distinct forms of knowledge. Bruce Seely argues that this emphasis is due in part to the close connection between American historians of technology and engineering education ("SHOT, the History of Technology, and Engineering Education," *Technology and Culture* 36, 1995, no. 4: 739–772, esp. p. 771). The classic article by Layton is "Mirror-Image Twins: The Communities of Science and Technology in 19th-Century America," *Technology and Culture* 12 (1971), October, pp. 562–580. For a summary of the science-technology discussions, see John M. Staudenmaier, *Technology's Storytellers: Reweaving the Human Fabric* (MIT Press, 1985), pp. 83–120.

16. From the perspective of practicing engineers, "engineering science" can be read as an oxymoron. From the perspective of a historian of technology, however, it represents the latest in a long series of efforts on the part of engineers to stake out epistemological turf. As early as the eighteenth century, when the engineering profession was first

taking shape in France, engineers were trying to establish a middle ground of knowledge between abstract science on the one side and practicing artisans on the other. They were not *savants* but neither were they mechanics. Their role was to connect the two: science provides the base that engineers connect to practice. The terrain of engineering, then, was what Diderot defined as "innovative public technological knowledge." See Ken Alder, "French Engineers Become Professionals; or, How Meritocracy Made Knowledge Objective," in *The Sciences in Enlightened Europe*, ed. W. Clark et al. (University of Chicago Press, 1999), p. 104.

17. As displayed in the MIT Museum's "Mens et Manus" exhibit, opened in June 2001.

18. Helga Nowotny, private communication, March 8, 2001.

19. Peter Galison, *Image and Logic: A Material Culture of Microphysics* (University of Chicago Press, 1997), pp. 46, 48.

20. Linda G. Griffith-Cima, "A Modest Proposal for Biomedical Engineering Education," *MIT Faculty Newsletter* 9 (1996), no. 2, pp. 17, 15, 14.

21. On biology's convergence with technology, see Castells, *The Rise of the Network Society*, pp. 29, 39, 54–59.

22. Ibid., p. 29.

23. When he became Dean of Engineering, Gordon Brown appointed a young new department head, Charles Miller, who pioneered in introducing computing into civil engineering. Miller had made his mark in the late 1950s by originating the Digital Terrain Model (DTM) concept, which replaced analog terrain data (in the form of contour lines or cross-sections) with terrain data in a numerical form which could be processed by a high-speed digital computer. Shortly after, Miller invented Coordinate GeOmetry (COGO), "the first problem-oriented language, making it possible for end users to communicate directly with the computer using the nomenclature and geometric building blocks of CE practice." Together DTM and COGO became the core of the Integrated Civil Engineering System, "the forerunner of numerous computer-aided design systems in use today," and a major research and development effort at MIT in the early 1960s. See "Readers' Notes," *Civil and Environmental Engineering at MIT* 14 (2000), no. 3, p. 8; "Comings & Goings," *Civil and Environmental Engineering at MIT* 13 (1998), no. 1, pp. 9–10.

24. Rafael Bras, "Note from the Department Head," *Civil and Environmental Engineering at MIT* 13 (1998), no. 1, p. 1.

25. L. L. Bucciarelli, H. H. Einstein, P. T. Terenzini, and A. D. Walser, "ECSEL/MIT Engineering Education Workshop '99: A Report with Recommendations," *Journal of Engineering Education* (April 2000), p. 141.

26. The sequence of action-oriented verbs may have been inspired by the military slogan OODA, meaning "observe, orient, decide, act."

27. From ESD Mission Statement, Endicott House retreat, January 2001.

28. I have been told that the distinction between engineering and Engineering was preceded in the 1980s by a similar distinction between manufacturing and Manufacturing. This earlier formulation appears to have faded from use, but "big E engineering" is still a commonly heard phrase.

29. Nancy Mulford, "Toward an Engineering Theory of Life," *MIT Report* 29 (2001), no. 2, p. 2.

30. Weatherall, "Engineering Education and Practice," p. 2.

31. *Engineering Education and Practice in the United States: Foundations of Our Techno-Economic Future* (National Academy Press, 1985). See also Weatherall, "A New Breed of Engineering Student," p. 2.

32. Weatherall, "A New Breed of Engineering Student," pp. 1–2.

33. McKinsey & Company, 1999 Engineering Alumni Career Decisions Survey, Overall vs. MIT findings (January 2000), pp. 3, 7, 8, 10.

34. Weatherall, Engineers as Supervisors and Managers, p. 7.

35. According to June 2000 data from Christopher Pratt, who succeeded Weatherall as director of MIT's Office of Career Services, 15% of the employers recruiting MIT graduates were from consulting companies and 11% from accounting, finance, business or financial services, and investment banking ("Career Development for Students: Programming and Financing a System," presentation material, June 16, 2000).

36. Data from Office of Career Services and Preprofessional Advising, "Changing Character of Employers Recruiting at the Careers Office, 1983–1995." The McKinsey study shows that over the past

20 years two-thirds of MIT engineering graduates ended up working in the field of their undergraduate degree, 13% in another technical/engineering field and 14% in a nontechnical/engineering field. (The numbers for the overall respondents are similar for the number working in the field of their undergraduate degree; 18% in another technical/engineering field; and 9% in a nontechnical/engineering field.) The number of graduates who migrate into nontechnical/engineering fields is therefore not overwhelming, but neither is it trivial (p. A47). The trend toward entrepreneurship in the 1999 McKinsey survey of engineering graduates is more pronounced among MIT graduates, though not radically so.

37. Castells, *The Rise of the Network Society*, pp. 217, 282. See his summary of post-1970 developments in chapter 4, "The Transformation of Work and Employment: Networkers, Jobless, and Flex-timers."

38. Killian, "Proposal for the Appointment of a Faculty Committee to Study Our Undergraduate Educational Program," memorandum to Dr. Compton, August 5, 1946 (MIT Archives, AC 4, Box 57, file 12).

39. *Report of The Committee on Educational Survey* (MIT Press, 1949), p. 39.

40. Ibid., p. 42.

41. Another doctoral degree in the humanities is offered by the History, Theory, and Criticism program in the Department of Architecture.

42. A quick analysis of MIT's requirements reveals the basic structure of the curricular demands. First, there is the great divide down the middle: half of MIT's requirements are General Institute Requirements (GIRs), which all students must fulfill, and half are departmental requirements, which are required of students who major in that department. Two-thirds of MIT's students major in some engineering department, so for most undergraduates half their schedule is devoted to GIRs and half to an engineering major. The GIRs in turn are also split into two equal parts: half are science requirements (the "science core," including two terms of math, two of physics, one of chemistry, and one of biology) and half are courses in humanities, arts, and social sciences (eight terms required in all, so that MIT students normally take one class each term in humanities, arts, and social sciences). The overall curriculum, then, could be shown on a pie chart as one-half departmental subjects, one-fourth basic science subjects, and one-fourth subjects in humanities, arts, and social sciences. This aspect of the MIT curriculum was put in place in the 1960s and

has been modified but not substantially changed since then. It represents a balance of competing demands among various faculty pressures, but most of all it represents a balance between the technical core of engineering education—represented by the departmental requirements—and two complements of that technical core: basic science as the essential foundation for engineering, and humanities, arts, and social sciences as essential "broadening" of engineering. The balance is at once delicate and robust. Nearly every faculty member would love to change this "great compromise" in a way that gives more attention to his or her discipline or passion (usually but not always the same). And nearly every faculty member realizes that trying to alter the "great compromise" would cause an enormous uproar that would be hard to contain or to manage productively. (The expression "great compromise" is used here with considerable irony, as it refers to an episode in American history where the stakes were much higher: the 1787 decision of the Constitutional Convention to draft a plan for a two-house legislature, with a House of Representatives apportioned according to the population of each state and a Senate in which each state would have the same number of votes.)

43. According to the historian Ken Alder, engineering education has long featured subjects that impart authority and legitimacy to engineers rather than serving any clearly useful purpose. In eighteenth-century French engineering education, mathematics "proved an alluring measure of merit" by providing a supposedly non-discriminatory standard for judgments about students, but its "functional contribution . . . to the practical success of the late-eighteenth-century artillery engineers . . . was largely illusory. Its real contribution was social and pedagogical." (Alder, "French Engineers Become Professionals," pp. 110, 116)

44. Christopher Pratt, Career Development for Students, handout.

45. *Report of the CES,* pp. 8–9, 20.

46. Ibid., p. 89.

47. Bucciarelli et al., "ECSEL/MIT Engineering Education Workshop '99," p. 141.

48. The report quotes a number of engineering faculty members who say quite explicitly that the reason for teaching design to engineering students has little or nothing to do with professional practice. In the words of a chemical engineer from MIT: "A look at what our students do after they leave MIT suggests that relatively few will ever design a [traditional refinery] process for profit. Given this, the

primary purpose of design in the curriculum is to assist students in developing an ability to solve problems, which will be very useful no matter how they proceed professionally." (p. 142)

49. John Henry Newman, "The Idea of a University" (1852), in *Victorian Prose*, ed. F. Roe (Ronald Press, 1947), p. 186.

50. Thomas P. Hughes, "Transdisciplinary Engineering, or *Gesamtingenieurkunst*," in "Memorials: Elting Morison, 1909–1995," *Technology and Culture* 37 (1996), no. 4, p. 879.

51. Hughes, personal communication to Larry Cohen of MIT Press, August 6, 2001.

52. Bruno Latour, seminar, MIT Science, Technology and Society Program, April 18, 2001. See also Bruno Latour, "Regeln für die neuen wissenschaftlichen und sozialen Experimente" (plenary lecture, Darmstadt Colloquium, 2001), p. 11.

53. Alder, "French Engineers Become Professionals," p. 124.

## Chapter 3

1. At MIT, 'staff' (widely used as a collective singular noun) normally refers to non-faculty employees, although faculty members are technically staff in many contexts.

2. Janet Snover, "Reengineering Is Over but Change Is Not," *MIT Faculty Newsletter* 12 (1999), no. 2, pp. 18–19.

3. Private email, July 8, 2001. This "MIT-like" flavor is not entirely coincidental. James Champy, the management consultant who co-authored *Reengineering the Corporation* with Michael Hammer in 1993, graduated from MIT in 1965 and later became a member of the MIT Corporation.

4. Snover, "Reengineering Is Over but Change Is Not," pp. 23, 26–27. When Harvard announced its multi-year project to replace central financial and human resource information systems in 1996, Project ADAPT, it estimated costs at around $50 million, "an amount that is comparable to what other universities have committed for similar efforts." *Harvard University Gazette* 91 (1996), no. 30, pp. 1, 10. By fiscal 1999, annual savings at MIT attributed to Reengineering were estimated at $15 million annually, enough to pay off the one-time costs of Reengineering by the end of fiscal year 2001.

5. Castells, *The Rise of the Network Society*, pp. 59–61.

6. Snover, "Reengineering Is Over but Change Is Not," p. 1.

7. See Agatha C. Hughes and Thomas P. Hughes, eds., *Systems, Experts, and Computers: The Systems Approach in Management and Engineering, World War II and After* (MIT Press, 2000); also Thomas P. Hughes, "The Spread of the Systems Approach," in *Rescuing Prometheus* (Pantheon, 1998), pp. 141–195.

8. Constance Perin, "Making More Matter at the Bottom Line," in *Corporate Futures: The Diffusion of the Culturally Sensitive Corporate Form*, ed. G. Marcus (University of Chicago Press, 1998), p. 86. Frederick Winslow Taylor's techniques for increasing workplace efficiency became a global management fad in the interwar years.

9. Most requests were declined, but MIT has established partnerships with other institutions (notably the University of Singapore and Cambridge University) and is even setting up what in the business world would be called a spinoff (MediaLab/Dublin).

10. Peter Marris, *Loss and Change* (Pantheon, 1974), pp. 2, 4.

11. Ibid., p. 4.

12. Ibid., p. 6.

13. Ibid., p. 21.

14. Thomas Parke Hughes, "The Evolution of Large Technological Systems," in *The Social Construction of Technological Systems: New Directions in the Sociology and History of Technology*, ed. W. Bijker et al. (MIT Press, 1987), pp. 76–80.

15. Castells, *The Rise of the Network Society*, p. 90.

16. Helga Nowotny and Ulrike Felt, *After the Breakthrough: The Emergence of High-Temperature Superconductivity as a Research Field* (Cambridge University Press, 1997), p. 23, quoting Jean-Claude Derian, *America's Struggle for Leadership in Technology* (MIT Press, 1990).

17. Ruth Schwartz Cowan, *More Work for Mother: The Ironies of Household Technologies from the Open Hearth to the Microwave* (Basic Books, 1983).

18. Snover, "Reengineering Is Over but Change Is Not," p. 1.

19. Wiebe E. Bijker, *Of Bicycles, Bakelites, and Bulbs: Toward a Theory of Sociotechnical Change* (MIT Press, 1995), p. 281. In subsequent work, Bijker and his colleagues have moved on from this "bracketing" of technological determinism (his expression, used in a

private email dated February 18, 2000) in order to address it in a more conceptually complex way.

20. See the discussion of "hard determinism" in Leo Marx and Merritt Roe Smith's introduction to their anthology *Does Technology Drive History? The Dilemma of Technological Determinism* (MIT Press, 1994), p. xii.

21. Castells, *The Rise of the Network Society*, p. 5.

22. Snover, "Reengineering Is Over but Change Is Not," pp. 23–24.

23. Of course, integration did not necessarily mean having an all-SAP environment. In some local areas, and even in some parts of the central administration, staff argued for other software that would be more robust and user-friendly in getting the work done, and that could easily be tied into SAP-based systems. Generally, however, the bias was toward an all-SAP environment.

24. Martin Bauer, "Resistance to New Technology and Its Effects on Nuclear Power, Information Technology and Biotechnology," in *Resistance to New Technology: Nuclear Power, Information Technology and Biotechnology*," ed. M. Bauer (Cambridge University Press, 1995), p. 19.

25. Robert Friedel is currently studying medieval cheesemaking. David Jardini ("Out of the Blue Yonder: The Transfer of Systems Thinking from the Pentagon to the Great Society, 1961–1965," in *Systems, Experts, and Computers*, ed. Hughes and Hughes, pp. 317–319) analyzes the "magnum opus of the RAND Corporation's 'scientific' defense analysts, *The Economics of Defense in the Nuclear Age*" (1960) as a manifesto intended to provide executives with "'scientific' decision tools that replaced inherently biased *experience* with rigorous, quantitative *analysis*."

26. Jardini ("Out of the Blue Yonder," p. 342) notes that one of the strongest proponents of the analytical systems approach, Robert McNamara, "made clear in his 1995 book, *In Retrospect*, [that] the management and decision-making processes adopted by him and his staff contributed directly to the military debacle in Vietnam."

27. The concept of society resisting and correcting market forces is central to Karl Polanyi's book *The Great Transformation* (Beacon, 1957 [1944]).

28. Jean-Luc Godard, quoted by Richard Brody, "Profiles: An Exile in Paradise," *New Yorker*, November 20, 2000, p. 64.

29. Rich Garcia, "IS Helps to Implement Student Information Policy," *I/S* [MIT publication] 16 (2000), no. 1, p. 6.

30. N. Katherine Hayles, "The Materiality of Informatics," *Configurations* 1 (1993), no. 1, pp. 150–153.

31. Perin, "Making More Matter at the Bottom Line," pp. 85–86.

32. For example, the Science Council, which Vest did much to help launch in 1994, has been a factor in the rise in federal funding, most dramatically for the National Institutes of Health (from $11 billion in 1994 to $17.8 billion in 2000) and the National Science Foundation (from $3 billion to $4 billion in the same period).

33. "Innovations in University/Industry Alliances: An Interview with Karl Koster," *MIT Report* 29 (2001), no. 3, pp. 2–3.

34. Just under a year after Scott Krueger's death, a grand jury investigation indicted the local chapter of the Phi Gamma Delta fraternity, but not individual students or administrators, nor the national fraternity, nor MIT. This indictment was important as a statement of responsibility, but largely ineffectual as a practical matter as the local chapter had already disbanded. Two years after the grand jury reported, MIT and the Krueger family arrived at a mediated settlement of $6 million (including $1.25 million for scholarships) and a public letter of apology from President Vest.

35. Sarah H. Wright, "'Dramatic' Changes at MIT since 1997 Cited by Students," *Tech Talk*, Wednesday, September 20, 2000, pp. 1, 8. The quote is from Damien Brosnan, president of the Interfraternity Council.

36. Wright, "'Dramatic' Changes at MIT," quoting Matt McGann.

37. Beck, *Risk Society*, pp. 19–21.

## Chapter 4

1. Hannah Arendt, "The Greek Solution," in *The Human Condition*, pp. 171–178.

2. There are five Schools at MIT: Engineering; Science; Architecture and Planning; Humanities, Arts, and Social Sciences (founded in 1950); and the Sloan School of Management (founded in 1951).

3. Task Force on Student Life and Learning [committee report], September 1998), p. 7.

4. In his widely read essay "Electronics and the Dim Future of the University" (*Science* 270, October 13, 1995, pp. 247–249), the social scientist Eli Noam argued that universities developed around collections of texts summarizing the knowledge of a society, scholars who studied those texts, and students who studied with the scholars.

5. The phrase "beloved community" is from Casey Nelson Blake, *Beloved Community: The Cultural Criticism of Randolph Bourne, Van Wyck Brooks, Waldo Frank, and Lewis Mumford* (University of North Carolina Press, 1990).

6. The place-based community is an entity that has a material basis in physical space. It is, in the language of phenomenology, a "thing," a visible and durable reality that provides a common point of reference for multiple individual perspectives. Hannah Arendt, who used the concepts of phenomenology to analyze the postwar "human condition," put it this way: "To live together in the world means essentially that a world of things is between those who have it in common, as a table is located between those who sit around it; the world, like every in-between, relates and separates men at the same time. . . . Only where things can be seen by many in a variety of aspects without changing their identity, so that those who are gathered around them know they see sameness in utter diversity, can worldly reality truly and reliably appear." (Arendt, *The Human Condition*, pp. 48, 52–53)

7. The final report of the Task Force included a separate section commenting on "the educational role played by MIT's dedicated staff members. Staff excellence is an integral part of today's MIT community. . . . The Task Force has met and worked with numerous MIT staff members in the past two years, and this has reinforced the Task Force's feeling that staff play a tremendously positive role in keeping the MIT community together." (p. 35)

8. Task Force report, p. 14.

9. The final report does include an important statement that "MIT is taking the lead in providing an education balanced between the practice of science and technology and liberal education. If successful, this will make MIT the model of a general education, giving the Institute a new competitive advantage." (p. 51)

10. Email, February 3, 1998. In the report of the Task Force, the section on "The Attributes of an Educated Individual" puts the idea in these words: "An educated individual possesses well-developed facul-

ties of critical and rational reasoning. She understands the scientific method and other methods of inquiry and hence is able to obtain, evaluate, and utilize information to pose and solve complex problems in life and work. To this end, she has a strong grasp of quantitative reasoning, and has the ability to manage complexity and ambiguity." (Task Force report, p. 17)

11. Raymond Williams, *Politics and Letters: Interviews with New Left Review* (Verso, 1981), p. 119.

12. Task Force report, p. 18.

13. Ibid., p. 33.

14. Ibid., p. 33.

15. Ibid., p. 7.

16. Ibid., p. 24.

17. Ibid., p. 31.

18. Ibid., p. 51.

19. Ibid., p. 53.

20. Ibid., p. 58.

21. Ibid., p. 58.

22. See Colin Bell and Howard Newby, *Community Studies* (Allen & Unwin, 1971). *Habits of the Heart* by Robert Bellah et al. (University of California Press, 1985) helped make "community" a part of the ongoing political discussion. *Habits of the Heart* lamented the dominance of individualism in American life and argued for the need to reconfigure our civic life so that commitment, community, and citizenship were more in balance. By the early 1990s "communitarianism" had become a popular term. By 1996, when a new edition of *Habits of the Heart* appeared, a lively discussion was ongoing as to the need to reinvigorate civil society. "Community" became a shorthand term used to describe a sought-after state of affairs that would restore the personal and social connections that had been so battered by neocapitalist individualism, especially since the 1960s. In the world of political theory, the idea of community is most closely associated with the theorists Michael J. Sandel and Alasdair MacIntyre. One of the most active leaders of the communitarian movement is the sociologist Amitai Etzioni. For a summary, see Steven Kautz, *Liberalism and Community* (Cornell University Press, 1995), pp. ix–xi, 1–6.

23. See Howard Rheingold, *The Virtual Community* (Addison-Wesley, 1993). These new communities are unbound from the soil. Instead of being located in a place and bound by external necessity, they are geographically distributed, connected by communications networks, and bound by shared interests. Since the ties are chosen, not given, people in these communities are surrounded by the like-minded: not like-minded in every detail, but sharing enough interests and world-view to define themselves as members. The major difference between a virtual community and a geographical one is not the absence or presence of "technology." In both cases, the life of the community depends on complicated technological webs. The most important difference is that the virtual community is self-selected for common consciousness. It is the ultimate Cartesian social group: it detaches mind from body, just as it detaches consciousness from place. It is a world we have made, a reflexive world, created by humans and endlessly reflecting human beliefs, desires, ideas. Goran Therborn's study *Science, Class and Society* (Tryck RevoPress, 1974) argues that the achievement of sociology as a discipline was to discover the category of the "ideological community," bound by shared norms and values. See Williams, *Politics and Letters*, pp. 113–114.

24. Nowotny and Felt, *After the Breakthrough*, p. 180.

25. This contradiction was classically expressed by John Henry Newman in a series of essays on "the idea of a university," written after he was invited, in 1851, to become rector of the proposed new Catholic University at Dublin. If he were to "briefly and popularly" summarize that idea, Newman wrote, he would say that a university is "a *Studium Generale*," or "School of Universal Learning": "This description implies the assemblage of *strangers* from all parts in one spot—*from all parts;* else, how will you find professors and students for every department of knowledge? And *in one spot*; else, how can there be any school at all?" Newman asserts that the learning must be universal, but it must be brought to ground in "a place of concourse, whither students come from every quarter for every kind of knowledge." It is a place that gathers "the best of every kind everywhere": a place where the professor becomes eloquent, where mind collides with mind, where the best art and the best ideas collect. "In the nature of things, greatness and unity go together: excellence implies a center." (John Henry Newman, "What Is a University?" [1854], in *Victorian Prose*, ed. Roe, pp. 181, 184–185.

26. Task Force report, pp. 13–14.

27. Anthony Giddens has analyzed the complicated geography of modern life as "a double-layered, or ambivalent, experience rather than simply a loss of community." He explains: "What happens is not simply that localized influences drain away into the more impersonalised relations of abstract systems. Instead, the very tissues of spatial experience alter, conjoining proximity and distance in ways that have few close parallels in prior ages." (Giddens, *The Consequences of Modernity*, p. 140)

28. Task Force report, pp. 8, 50–51.

29. Nowotny and Felt, *After the Breakthrough*, p. 180.

30. David Harvey, *The Condition of Postmodernity: An Enquiry into the Origins of Cultural Change* (Blackwell, 1989), p. 204.

31. Task Force report, pp. 36–39.

32. Ibid., pp. 41–45.

33. Ibid., p. 39.

34. Ibid., p. 41.

35. There is no need to appeal to Einstein here. The philosophical point was made much earlier by Newton's contemporary and adversary Leibniz, who asserted that without "things" there would be no time and no space: "I have more than once stated that I held *space* to be something purely relative, like time; space being an order of co-existences as time is an order of successions." (*Leibniz: Philosophical Writings*, ed. G. Parkinson, London: J. M. Dent, 1973, pp. 211–212, quoted in Castells, *The Rise of the Network Society*, p. 494 and note.)

36. Arendt, *The Human Condition*, pp. 230–231.

37. Martin Moore-Ede, *The Twenty-Four Hour Society: Understanding Human Limits in a World That Never Stops* (Addison-Wesley, 1993), p. 4.

38. Donn Welton, "World," in *Encyclopedia of Phenomenology*, ed. L. Embree et al. (Kluwer, 1997), pp. 736–737.

39. Thomas Nenon, "Life-World in Husserl," *The Encyclopedia of Philosophy Supplement*, ed. D. Borchet (Macmillan Reference, 1996), p. 305.

40. I therefore prefer the "strong" version of the lifeworld, as described by Fred Dallmayr: "In the strong version, the life-world functions no longer as a mere precursor of reason or as its relatively immature or embryonic modality, but rather emerges as an integral

dimension of thought, a dimension impinging powerfully on the status of rational or cognitive claims (not by nullifying them, but by changing their sense)." ("Life-World: Variations on a Theme," p. 26)

41. I am alluding to Henri Lefebvre's classic work *La Production de l'espace* (Anthropos, 1974).

42. "Letter to Faculty" signed by Robert L. Jaffe, Ahmed F. Ghoniem, and Steven R. Lerman, *MIT Faculty Newsletter* 13 (2001), no. 4, p. 5.

43. "Debugging 6.001," *The Tech* 120 (2000), no. 43, p. 4.

44. Harvey, *The Condition of Postmodernity*, p. 232.

45. The Faculty Club, which once had a bar and was open for lunch daily, has been closed to individuals for some years. It is available to MIT groups and committees on a rental basis. Many at the Institute still speak of its "closing."

46. Task Force report, p. 41.

47. Staffan Burenstam Linder, *The Harried Leisure Class* (Columbia University Press, 1970), pp. 60, 67.

48. Harvey, *The Condition of Postmodernity*, p. 227, quoting Marx, *Grundrisse* (Penguin, 1973), p. 173,

49. Nancy Folbre, cited by Ellen Goodman, "Our 'prisoner of love' problem," *Boston Globe*, May 10, 2001.

50. Ann Crittenden, *The Price of Motherhood: Why the Most Important Job in the World Is Still the Least Valued* (Metropolitan Books, 2001).

51. 1998 Higher Education Research Institute Faculty Survey, as reported by Lydia Snover.

## Chapter 5

1. "In 1997 women in the United States earned 20 percent of the nation's math and computer science degrees and 12 percent of its engineering degrees. . . . Globally, . . . women earn only 10–12 percent of all PhDs in engineering and natural sciences, compared to nonscience degrees where the percentage is almost 50 percent." (Metin Akay, "Women in Science—An Untapped Resource," *Boston Globe*, July 24, 2001.)

2. More precisely, 52% of the women on the faculty of the School of Engineering have children, according to 2000 data. According to 1989 data, 82% of the male faculty members in that school have children.

3. 1995 Higher Education Research Institute Faculty Survey, as reported by Lydia Snover.

4. See Ruth Schwartz Cowan, "Technology Is to Science as Female Is to Male: Musings on the History and Character of Our Discipline," *Technology and Culture* 37 (1996), no. 3, pp. 572–582.

5. I am alluding here to Tracy Kidder's classic (and, through no particular fault of Kidder's, very male-dominated) book *The Soul of a New Machine* (Little, Brown, 1981).

6. Neil Gershenfeld, as reported in Mulford, "Toward an Engineering Theory of Life," p. 2.

7. This reduction of innovation to gadgets is evident in the magazine *Technology Review*, which I am told has no formal ties to MIT but whose masthead proclaims that it is "MIT's Magazine of Innovation." The innovation it celebrates is entirely technological, and women are notably absent in its choice of columnists and in its stories.

8. Any discussion of the cultural transition from "prehistory" to "history" is bound to be highly speculative. See, e.g., Robert McElvaine, *Eve's Seed: Biology, the Sexes, and the Course of History* (McGraw-Hill, 2001).

## Chapter 6

1. These concluding remarks reflect post-September 11 conversations with David Mindell, Dibner Career Development Professor in MIT's Program in Science, Technology, and Society.

2. Richard Willing and Kevin Johnson, "Plot Likely Involved," *USA Today*, October 5, 2001.

# Index